MW01154430

TRAVEL BY DESIGN

SPATIAL INFORMATION SYSTEMS

General Editors

M. F. Goodchild
P. A. Burrough
R. A. McDonnell
P. Switzer

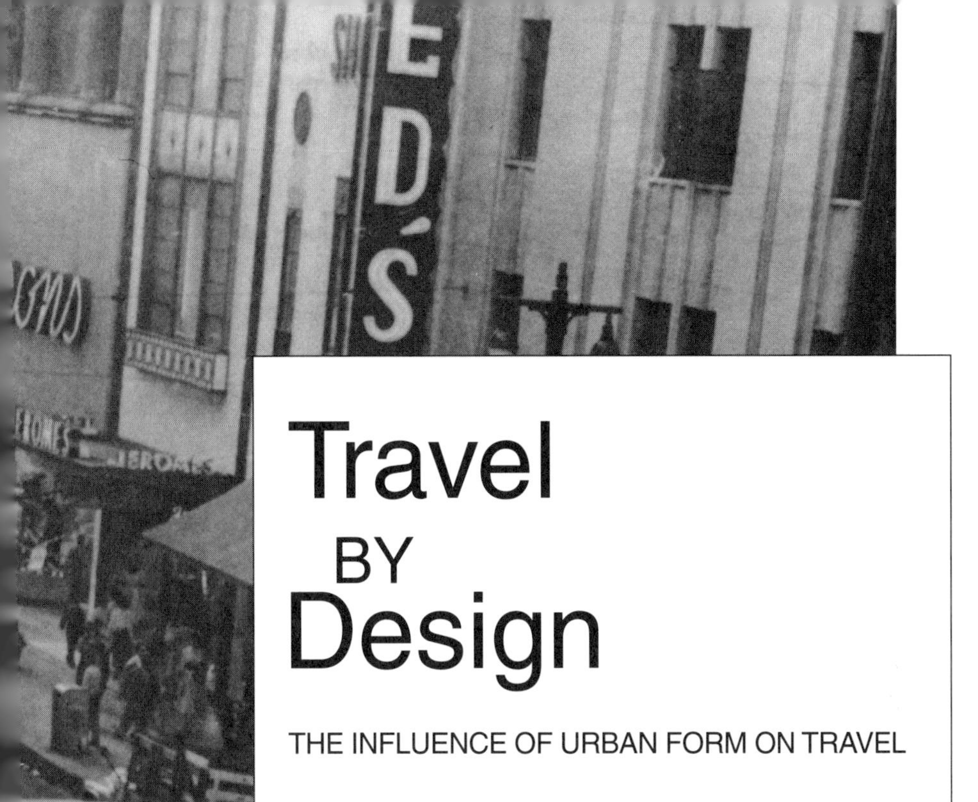

Travel BY Design

THE INFLUENCE OF URBAN FORM ON TRAVEL

Marlon G. Boarnet
Randall Crane

OXFORD
UNIVERSITY PRESS

2001

3129

OXFORD
UNIVERSITY PRESS

Oxford New York
Athens Auckland Bangkok Bogotá Buenos Aires Calcutta
Cape Town Chennai Dar es Salaam Delhi Florence Hong Kong Istanbul
Karachi Kuala Lumpur Madrid Melbourne Mexico City Mumbai
Nairobi Paris São Paulo Shanghai Singapore Taipei Tokyo Toronto Warsaw

and associated companies in
Berlin Ibadan

Copyright © 2001 by Oxford University Press, Inc.

Published by Oxford University Press, Inc.
198 Madison Avenue, New York, New York 10016

Oxford is a registered trademark of Oxford University Press.

All rights reserved. No part of this publication may be reproduced,
stored in a retrieval system, or transmitted, in any form or by any means,
electronic, mechanical, photocopying, recording, or otherwise,
without the prior permission of Oxford University Press.

Library of Congress Cataloging-in-Publication Data
Boarnet, Marlon, 1961–
Travel by design : the influence of urban form on travel
Marlon Boarnet, Randall Crane.
p. cm. — (Spatial information systems)
Includes bibliographical references
ISBN 0-19-512395-6
1. Urban transportation—Planning.
2. Trip generation—Mathematical models.
I. Crane, Randall. II. Title. III. Series.
HE305 .B63 2000
388.4′91314—dc21 99-053831

9 8 7 6 5 4 3 2 1

Printed in the United States of America
on acid-free paper

For

Barbara, Harrison, Jacob, and Michela (M.B.)

&

Max Roberto, Peter Alonso, and Marta Regina (R.C.)

Acknowledgments

We thank many individuals and institutions for their assistance with this work. We gratefully acknowledge research jointly conducted with Nicholas Compin, Richard Crepeau, and Sharon Sarmiento. In particular, Mr. Compin co-wrote chapter 8, and Mr. Crepeau's contributions to chapters 3 and 5 and Ms. Sarmiento's contributions to chapter 5 were all critical. Further, the research assistance of Compin, Crepeau, Richard Daulton, Sherry Ryan, and Dru van Hengel was essential to moving this project along at various stages. Much of their and our time was in turn financed by the University of California Transportation Center via grants from the U.S. and California Departments of Transportation.

Earlier versions of this material were first presented at the annual conferences of the Transportation Research Board, the Association of Collegiate Schools of Planning, the American Planning Association, a TRED (Taxation, Resources, and Economic Development) symposium at the Lincoln Institute for Land Policy, and seminars at Ohio State University, Portland State University, U.C. Berkeley, U.C. Irvine, UCLA, and the University of Southern California. We are grateful to participants and formal discussants at these presentations for their feedback and continuing questions. In addition, many individuals provided critical advice at key stages of this work, including Alex Anas, Marta Baillet, Nancy Bragado, Jan Brueckner, Dan Chatman, D. Gregg Doyle, Gordon "Pete" Fielding, Peter Gordon, Paul Gottlieb, Peter Gordon, Elizabeth Deakin, Kara Kockelman, Charles Lave, Kenneth Small, Brian Taylor, Martin Wachs, and Mel Webber. We are particularly grateful for the detailed and thoughtful comments of Robert Cervero, Reid Ewing, Susan Handy, and Jonathan Levine on parts of the last draft. The end result would no doubt be better if we had followed all their suggestions.

Parts of this book were originally published in somewhat different form in *Access*, the *Journal of the American Planning Association*, the *Journal of Planning Education and Research*, the *Journal of Planning Literature, Transportation Research A, Transportation Research D: Transport and Environment*, and *Urban Studies*. We thank their publishers for permission to draw on that work here, and we especially appreciate the efforts of their referees and editors to improve earlier versions of these materials. The cover photograph and those introducing each chapter are from the Security Pacific Collection, Los Angeles Public Library. They are all of Los Angeles and, with the exception of the 1939 photo of Sunset Strip by Fred William Carter on page 2, the photographers are unknown. (The cover is downtown Los Angeles in November 1953, looking south on Broadway.)

Finally, we deeply thank our senior OUP editor Joyce Berry for her interest in, support of, and patience with this project. The efficient efforts of the OUP production editor, Lisa Stallings, are also gratefully acknowledged.

Contents

PART I INTRODUCTION

1: An Overview of *Travel by Design* 3
Urban Design and Transportation Planning 3
Travel by Design? 5
Will the New Designs Work? 11
Will the New Designs Be Implemented? 12
Urban Design in the Context of Transportation Policy 13

PART II TRAVEL BEHAVIOR

2: The Trouble with Traffic 17
The Theory of Social Costs 19
Air Quality Regulation and Automobile Emissions 20
Traffic Congestion 25
Neighborhood Quality of Life 29
Summary 30

3: Studies of Urban Form and Travel 33
The Influence of Transportation on Urban Form 33
The Influence of Urban Form on Travel 35
Hypothetical Studies 39
Descriptive Studies 44
Multivariate Statistical Studies 47
Summary 58

4: The Demand for Travel 61
A Behavioral Framework 61
The Generic Impacts of Design Features on Travel 68
Issues for Applied Empirical Work 72
An Empirical Strategy 75
Summary 79

5: An Empirical Study of Travel Behavior 81
Overview and Data 81

Base Models 87
Incorporating Choices About Residential Location: An Empirical Model 94
Incorporating Choices About Residential Location: Results 96
Summary and Discussion 103

PART III THE SUPPLY OF PLACE

6: Neighborhood Supply Issues 109
Neighborhood Supply 109
Market Failure and Government Failure 111
Planning Incentives 114
Policy and Neighborhood Supply 116

7: Transit-Oriented Planning 119
Transit-Oriented Planning 119
Evidence from Southern California 124
Interpreting the Evidence 131
Two Behavioral Models of Transit-Oriented Planning 132
Summary 143

8: A Case Study of Planning 147
Data Sources and Background 148
Transit-Oriented Development in the City of La Mesa 155
Transit-Oriented Development in the City of San Diego 160
Barriers to TOD Implementation in San Diego County 162
Summary 166

PART IV WHAT ROLE FOR TRAVEL BY DESIGN?

9: Lessons for Research and Practice 171
The Influence of the Built Environment on Travel 172
Neighborhood Supply Obstacles 174
Is Urban Design Good Transportation Policy? 175
Policy Implications: Incentive versus Outcome Regulation 177
Directions for Future Research 179

Appendix: Data and Data Collection Methods for Chapter 7, by Source 181

Notes 185
References 201
Index 221

PART I
Introduction

The idea that neighborhoods and cities can be designed to change travel behavior holds much currency these days. In a manner of speaking, it promises to kill two birds with one very attractive stone: reduce car use and increase the quality of neighborhood life generally by improving the pedestrian and transit environments, all in the form of pretty houses and friendlier, often nostalgic, streetscapes. This idea prompts several questions, three of which are explicitly addressed in this book: Can it work, will it be put into practice, and is it a good idea? In this chapter we outline our approach to each.

1 An Overview of *Travel by Design*

"A street is a street, and one lives there in a certain way not because architects have imagined streets in certain ways." (Culot and Krier, 1978, p. 42)

Urban Design and Transportation Planning

Travel is not a simple story.

Start with the trips people make from home to work, and then back home again. Each commute reflects choices of where to live, where to work, when to work, when to go home, how to get from home to work, and what side trips to make along the way. Each decision depends on the opportunities available, with those in turn explained by the characteristics, resources, and values of workers, their families, their employers, other travelers, and of course the built environment of sidewalks, streets, bus routes, and rail lines connecting home to work. Nonwork trips, the great majority of trips in modern times, entail even more finely detailed mosaics of people, places, and the variety of things one obtains, or hopes to obtain, by going somewhere. Travel is the outcome of a grand confluence of human and other factors, many systematic and many others not. It will never be fully understood.

But because travel poses numerous challenges, and opportunities, it would be good to understand more. Planning strategies to reduce traffic congestion and improve air quality continue to get prominent attention. Several increasingly influential efforts emphasize the potentially mitigating role of the built environment.

For example, a good deal has been made in recent years of the fact that people drive less and walk more in downtown San Francisco than in suburbs anywhere. Part of this observed behavior is no doubt attributable to the kinds of people living there, people who prefer and indeed seek out the many benefits—travel and otherwise—of a diverse, high-density, mixed-use environment. But many observers have also asked, quite reasonably, if it would not make sense to design suburbs and other neighborhoods to be more like downtown San Francisco, or more like whatever it is about those places that leads people to drive less. Perhaps then people in suburbs and elsewhere would drive less and walk more. And perhaps that would lead to improvements in traffic congestion, air quality, and other transportation problems associated with the automobile.

Would that work? And if so, which features of these designs are most effective, which least effective, and why? These questions strike us as very interesting and challenging, given the complexity of travel behavior, and we turned with great anticipation to the massive literature on the subject for answers. Our reading led to a surprising conclusion: *Very little is known regarding how the built environment influences travel, and there is little agreement on how to reliably learn more.*

So we wrote this book.

Certainly it is not intended to be the last word on the subject. Rather, our main goal is to explain just this much: If one wanted to know how the built environment influenced travel, in order perhaps to design communities with less driving, how would one go about it? Toward that end, we assess what is currently known about urban form and travel behavior and suggest how we can learn more.

The challenge facing transportation planners in the first part of this century was to design and build the infrastructure needed to support a new product, the car. The task was formidable, and road building was the biggest infrastructure project in industrialized nations throughout the twentieth century. It is estimated that as of 1990, roughly one third of the value of all public infrastructure in the United States comprised streets and highways (Gramlich, 1994).

The planners who designed and built these roads were physical, not social, engineers. They saw their task as building street and highway capacity to meet certain precisely specified vehicle flow and circulation objectives. They rarely sought to change urban form to influence travel patterns.[1] Instead, they took existing travel patterns as a given, designed a road system to meet current and projected demand, and constructed the system. For many years, with the exception of plans by Olmsted (1924) and a few others, the only link between urban design and automobile transportation was the neighborhoods, often low income, that were divided and paved over to accommodate new freeways.

The idea that transportation planners would not manipulate urban form was even formalized in the planning process that still dominates virtually all transportation projects in the United States and abroad. Almost every large transportation project starts with a projection of travel demand. How many persons must the system carry from one location to another? The standard answer is obtained from what is known as the "four-step method" of travel demand estimation.

This method first divides an urban area into several small (often less than one square mile) zones. Survey data are then gathered on how many persons live in each zone and work in each zone, and often characteristics of those residents and jobs. Using the four-step model, one can then construct traffic flows from zone to zone for the morning and evening commute. Those commute flows are modeled as functions of land use, such as residential density within zones, zone employment density, and other variables (for early work of this kind, see Mitchell and Rapkin, 1954; more recent references include Domencich and McFadden, 1975; Ortuzar and Willumsen, 1994; Garrett and Wachs, 1996). Importantly, land use, urban design, and other elements of the built environment, to the extent that they are represented in the four-step model, are used to predict trip flows from zone to zone. Transportation planners have rarely tried to manipulate the built environment to influence travel patterns.

Times change, however. A variety of compelling and increasingly influential planning strategies propose a radical departure from past practice. While they differ in many respects, these efforts share the goal of limiting automobile use and a belief that the best means for doing so is by reshaping urban form.

Travel by Design?

One of the most popular planning ideas of the 1980s and 1990s is the set of design concepts now collectively known as the New Urbanism. These ambitious efforts have accepted the challenge of rethinking the relationships among form, scale, and movement in modern urban environments. The most visible proponents have been architects, especially the Miami team of Andres Duany and Elizabeth Plater-Zyberk (1991, 1992), best known for their work on the community of Seaside, Florida—cast as the fictional town of Sea Harbor in the 1998 film *The Truman Show*—and Peter Calthorpe (1993), who is based in San Francisco and is the coauthor of the "pedestrian pocket" concept (Kelbaugh, 1989).[2] While the proposals and projects differ in many respects, they share an emphasis on establishing the sense of community that often is missing in newly developed neighborhoods, to be accomplished largely by mixing land uses and getting people out of their cars.

The popularity of these ideas is not surprising. It is easy for complaints about cars and neighborhood form to get our attention. Cars pollute the air and traffic congestion eats up our time, whatever the overall value of the automobile in a mobile society. Cars likewise tend to monopolize the "public space" of the street, which had always been a key element of the social fabric (Appleyard, 1981; Lynch, 1981; Kostof, 1991, 1992; Southworth and Ben-Joseph, 1995, 1997). Thus, even freshly built neighborhoods seem to lack charm, and perhaps in certain respects they lack functionality as well. In place of the friendly front porch of older times, for example, the main exterior feature of a new house is most often the garage door (Southworth and Owens, 1993). It would be difficult to maintain that many new developments form true neighborhoods in the social sense, as there is little in their physical surroundings to link their residents privately or publicly aside from broad streets and the common architectural theme of their homes.[3]

The proposals for many less auto-dependent designs are also quite amiable. They are easy on the eyes, for one thing, and self-consciously familiar. The designers realized that to coax people to walk more, neighborhoods must be more pleasant to walk through and destinations must be closer. A major contribution of the path-breaking work in this field was to recognize that the prototypical New England or Southern small town, and "village" themes generally, fit the bill quite well.[4] Some survey evidence suggests that many suburbanites prefer to live in such towns, or at least in communities resembling them (Inman, 1993), and that effect is more or less what the New Urbanist and Neotraditional plans try to deliver: a physical environment inviting neighborhood interaction, rather than obstructing it, and land-use and street patterns permitting more travel by foot, all in a manner and appearance consistent with our collective sense of the traditional small town.[5] In principle, the new designs thus accept, rather than challenge, how many people would like to live.

The impacts of such thinking on professional practice have, roughly speaking, followed three lines. One is principally "architectural" in the sense that design and scale elements dominate. The community of Seaside (figs. 1.1–1.3), for example, is justly noted for the clapboard beauty of its homes, its white picket fences, and its weathered old-town feel, though it is barely fifteen years old (Mohney and Easterling, 1991). The look is sensitive to local context, however. The newer and larger

Figure 1.1. Aerial photo of Seaside. (Photo by Michael Moran)

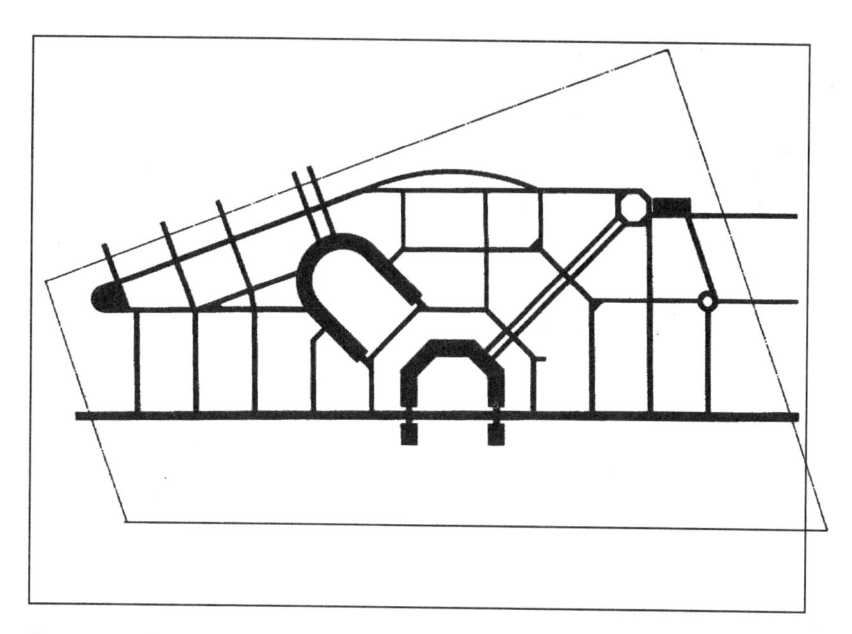

Figure 1.2. Street network, Seaside master plan. (Courtesy of Duany/Plater-Zyberk)

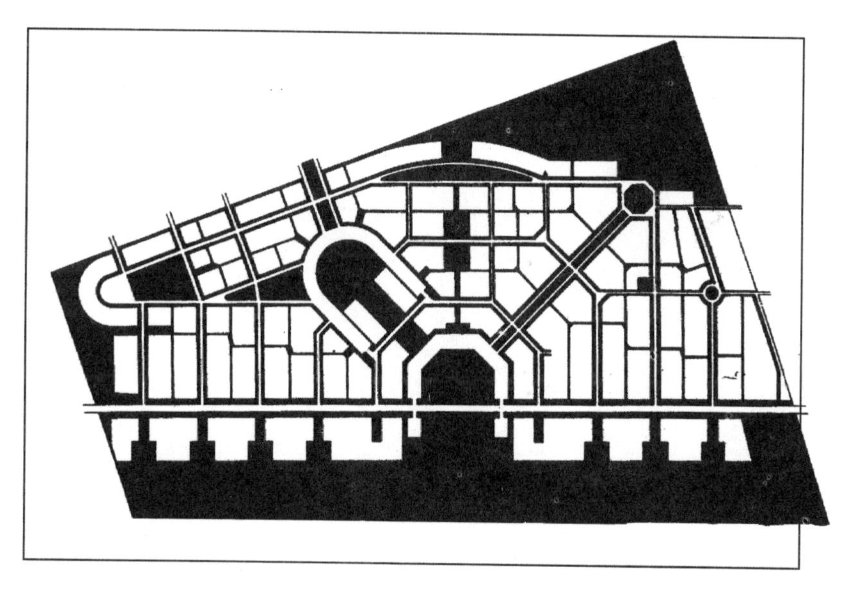

Figure 1.3. Pedestrian network, Seaside master plan. (Courtesy of Duany/Plater-Zyberk)

Duany/Plater-Zyberk project of Kentlands, in Gaithersburg, Maryland, is based on the mid-Atlantic look and feel of Annapolis and Georgetown.

In his writing and, to a lesser extent, in designs such as the Sacramento, California, area development of Laguna West, Calthorpe (1993) has stressed the importance of bringing human scale not only to individual housing tracts, but also to the linkages between residential and commercial activities (figure 1.4). The renewed emphasis on front porches, sidewalks, and common community areas as spatial focal points, as well as the half-mile-wide "village scale" of each community, are the most visible examples of such links. This last feature is strongly reminiscent of the "neighborhood unit" approach to planning first popularized in the 1920s and 1930s (Perry, 1939; Dahir, 1947; Banerjee and Baer, 1984).

Another area of influence is social theory. Some proponents of the new design strategies are the latest in a long line of social commentators who have looked with dismay at post–World War II suburbs. The complaints are many. Suburbs drain the middle class and their fiscal resources from central cities, leaving them warehouses of the poor (Downs, 1994). Suburbs isolate persons who no longer interact with others in public places. Instead, according to some observers, the typical suburban resident drives alone to work, private health clubs, movie theaters, and enclosed shopping malls, where any social interaction is only a shadow of the varied public life of major cities (a related and compelling discussion of many such issues is found in Waldie, 1996).

The third major area in which these designs have found popular acceptance is transportation policy. Public complaints about automobile

Figure 1.4. Laguna West, California. (Photo by and courtesy of Calthorpe Associates)

congestion and air quality have left planners intensely receptive to new ways of reducing car use, yet their options are limited. The cost of mass transit is ballooning out of proportion to expected benefits, and conventional transportation planning strategies have not changed the affection most people continue to feel for their cars (Giuliano, 1989; Deakin, 1991; Wachs, 1993a, 1993b). Fundamental change in land-use patterns is seen by some as a potentially more promising tool, and this idea has found its way into an increasing number of public planning and policy documents aimed at improving air quality or congestion by means of land-use/transportation linkages (San Diego, 1992; Los Angeles, 1993; San Bernardino, 1993).

As a solution to all of these problems, these designers propose to build what they regard as the smaller, more lively, more humane communities that in many ways evoke a bygone era. Because the car, in the view of many of the new urban designers, is the lifeblood of a flawed urban form, a central tenet of the new designs is the taming of the car (e.g., Warren, 1998). While the designers of the new communities apparently did not set out to make transportation policy, they found it difficult to avoid. Early on, they realized that transportation is vitally important to their broader design goals. Andres Duany was recently quoted in *Consumer Reports* (1996) as saying that the transportation elements of the New Urbanism are perhaps its most important.

Though the architects and planners promoting these ideas are usually careful to emphasize the many ingredients necessary to obtain desired results—the straightening of streets to open the local network, the "calming" of traffic, the better integration of land uses and densities, and so on—a growing literature and number of plans feature virtually any combination of these elements as axiomatic improvements. The conclusion that auto travel will decrease in more compact and gridlike land-use developments is so appealing that it has been reported as a virtual fact in almost all discussions of the new design principles.[6]

The result has been striking. In a few years, the New Urbanism has achieved prominence in the jargon of mainstream planners. Trade journals report on the latest designs and proposals. The Institute of Transportation Engineers has developed street design guidelines for New Urbanist developments (Institue of Transportation Engineers, 1997). Early neotraditional communities, such as Seaside, Florida, Kentlands, Maryland, and Laguna West, California, are discussed widely in scholarly articles, government reports, and informal discussions among planners. Cities such as Portland, Oregon, have promoted the new concepts, and some are actively banking on those designs to contribute to transportation goals (e.g., 1000 Friends, 1996). Transportation plans now often feature a prominent place for urban design.

In place of the traditional concern with providing road capacity, the new transportation efforts focus on the undesirable side effects of automobile use, including air pollution and traffic congestion. Furthermore, they focus both implicitly and explicitly on nonwork travel. The dense, mixed-use neighborhoods with public spaces are often designed to encourage walking trips to shops, schools, day care, and entertainment. This focus on nonwork travel is consistent with the growing importance of nonwork trips in daily travel. Over three-quarters of all urban trips are for nonwork purposes (NPTS, 1993). In many urban areas, traffic congestion is no longer strictly a rush hour phenomenon (Gordon, Kumar, and Richardson, 1989a).

The goal of using urban design as transportation policy has strong antecedents in the jobs-housing balance debate that first achieved prominence in the 1980s (Cervero, 1989b; Giuliano, 1991). Proponents of jobs-housing balancing argue that suburban communities are fundamentally unbalanced—that residences are concentrated in some neighborhoods while jobs are clustered in office parks. According to advocates of jobs-housing balancing, these strictly separated land uses make automobile travel a requirement for almost all suburban commuters. On the other hand, if planners designed communities with mixed uses, placing some jobs near residences, perhaps many more persons would be able to walk, use transit, or carpool to work. This focus on mixed land uses designed to reduced automobile travel foreshadowed similar themes in current design principles. From a transportation perspective, the new plans took that idea and applied it to all travel, nonwork travel included.

Like any bold new idea, the use of urban form to solve traffic problems raises many questions. This book focuses on three:

First, can it work? If we build cities and their suburbs differently, will their residents drive less? Our primary purpose here is to clarify the appropriate means for answering such a question, that is, to better understand how urban form generally influences how people travel. We also analyze both earlier studies and new data on observed travel behavior. In short, while many regard the influence of urban form on driving as either obvious or proven, we conclude it is neither. On a more optimistic note, we also clarify the circumstances under which urban design can potentially change travel behavior.

Second, can and will the new plans be implemented? There exist many design proposals for communities with these features, but precious few actual developments. Why? One explanation is that land use planning in the United States is overwhelmingly the domain of municipal governments, but the extent to which cities want to, and indeed do, plan land use toward transportation ends is all but unknown. Understanding the incentives and behavior of those local governments is vital if the new plans are to move from idea to practice. We place this issue within the larger context of the government regulation of neighborhood types, and present a systematic analysis and case study of one specific development strategy: transit-oriented development. Municipal incentives appear to be a key factor in explaining which kinds of neighborhoods are built. So, yes, the new designs *can* be built, but a deep understanding of the motives of local land-use authorities would seem to be key.

Finally, is this type of strategy a good idea? What are the pros and cons? Is driving less the solution to the problems associated with car use? How do the new urban designs compare to other possible policy responses to transportation problems? We defer these issues until after the first questions about workability and implementation have been discussed because understanding both the gaps in our knowledge and the implementation challenges help inform an assessment of the policy wisdom of using urban design as a transportation planning tool. The question is, in part, how well direct policy interventions such as pricing compare with more indirect regulatory policies such as urban design. In closing, we argue that the answer depends on the local policy context, yet we also identify several "rules of thumb" that may provide guidance in many situations.

We deal with each question in detail in this book and introduce them further in the remainder of this chapter.

Will the New Designs Work?

There is a mismatch between what we know about travel behavior and what we need to know to evaluate the transportation goals of urban designers.

In addressing this gap, a useful starting point is to view the transportation goals of urban design much as one would view any transportation strategy. This implies that the transportation benefits of urban designs should be weighed against their costs. This benefit-cost test is not controversial within the context of transportation projects or policy analysis more generally (but for an alternative approach, see Southworth, 1997).[7] Within that evaluative framework, we turn to the measurement of transportation benefits.

As noted above, the intuitive appeal of the new designs is strong. The idea is as simple as arguing that if we build communities where walking is more possible, people might walk more. Similarly, if we cluster shopping near rail or bus nodes, maybe transit will be used more for shopping trips. On a more general level, if we build communities the way they were built before the automobile, and otherwise encourage preautomobile modes of travel, it seems sensible to think that persons will drive less.

This intuition is buttressed by some longstanding relationships between urban form and travel behavior. First, transit ridership is generally higher in more dense cities (Pushkarev and Zupan, 1977). Second, neighborhoods with more destinations, such as workplaces and shopping, somewhat naturally are the terminus for more trips. Third, persons are willing to walk only very short distances in urban areas—often not more than a quarter mile (Untermann, 1984). Given all this, how could building dense, mixed-use neighborhoods that put trip destinations very close to residences *not* reduce automobile travel?

This is a fair question, to be sure, but the evidence appears mixed just the same. Many studies fail to find a clear link between the built environment and travel behavior at the margin. Others suggest that some of the ideas incorporated in the new urban designs might be associated with more automobile travel, rather than less. To better understand where and why such results differ, we review the existing literature in chapter 3. That review raises more questions than answers. Often the data are poorly suited to a rigorous study of the issues at hand; in other cases the framework used to assess the data is somewhat ad hoc and thus the results are difficult to either interpret or generalize.

How should studies relating urban design to travel proceed? What is the nature of the travel behavior that interests us, and how is it connected to the characteristics of the physical environment? In chapter 4 we establish two key parts of our research strategy. There we develop a theoretically consistent choice framework for analyzing how urban form influences travel behavior. We argue that the relationship between transportation behavior and the built environment is not as simple as is often assumed. Changes in urban design can influence automobile travel in ways that are hard to anticipate. As one example, shortening trip distances may promote walking, but they might also increase the number of trips taken by car. People may decide to shop more often and they may well continue using their cars to do so. In general, it is hard to say how specific urban design characteristics will affect travel. Put another way, these are empirical questions that can be settled only by analyzing data from a particular place at a particular point in time.

We then go on to suggest the form of the empirical tests in chapter 4. We contend that many of the deficiencies in the literature can be overcome by systematically isolating the separable influences of urban design characteristics on travel and then properly analyzing individual-level data. The first part of chapter 4 clarifies which results follow directly from alternative land-use arrangements and which may or may not; the latter part identifies the specific hypotheses to be tested against the data.

In chapter 5 we test these hypotheses in a variety of ways for two different sets of data. In addition to comparing results for two urban areas, chapter 5 explores the implications of alternative behavioral assumptions regarding travel costs. The measured influence of land use on travel behavior is shown to be sensitive to the form of the empirical strategy, the form of the data, and the specific community. Perhaps more important, the purpose of chapter 5 is to illustrate how empirical studies can be conducted in a theoretically and statistically consistent manner.

Will the New Designs Be Implemented?

The theory and empirical evidence in chapters 3–5 constitute a behavioral analysis of how urban design can influence travel behavior. Yet policy analysis cannot stop there. Even where urban designs hold the promise of achieving their transportation goals, will they be implemented?

Community design and building span the arenas of private land development and government regulation. For most developments to be feasible, they must have the promise of being both marketable and consistent with local land-use regulations. This suggests two questions: Will persons want to live in these communities, and how will local governments react to those development proposals?

There is some evidence on market potential which suggests that moderately dense, mixed-use developments can appeal to some, but not all, segments of the suburban housing market (Fulton, 1996). Yet there is only scant evidence on how government regulations enable or constrain such developments (Levine, 1998).

Land-use regulation in the United States is almost the exclusive province of local governments. Air quality and congestion problems span municipal borders. Thus, any urban design or land-use solution to regional air quality and traffic problems requires intergovernmental coordination. This point has been almost completely overlooked in the context of the transportation goals of urban designs. How do local governments tend to respond to regional goals when they relate to local land use? From a municipal perspective, what are the perceived advantages and disadvantages of alternative urban designs? Is there a mismatch between local land-use goals and regional transportation needs? These are the topics examined in chapters 6, 7, and 8.

Chapter 6 sets the stage for this analysis by placing municipal land-use regulation within the context of governmental behavior. In chapters 7 and 8 we present evidence on the experience of southern California regarding municipal incentives toward one particular design initiative: the development of commuter rail station areas for transit-supportive housing, a component of transit-oriented development. As in preceding chapters, our emphasis is on how to study these questions, as well as generating new empirical results. We find that, in this instance, for this type of design, implementation faces many obstacles. This in turn requires a discussion of where and how transit-oriented development strategies might best address those obstacles.

Urban Design in the Context of Transportation Policy

Analyzing the transportation planning potential of urban design requires a policy context. How does the form of the built environment compare with other transportation planning tools? Are the new urban designs a promising means to reduce traffic congestion and improve air quality when compared with other options? Building on our analysis of the effect of the built environment on travel behavior and the implementation prospects for new urban designs, we compare the transportation impacts of urban design policies to those of other policies in chapter 9. The outcome is a comprehensive policy assessment of the transportation planning element of urban design—something that to date has been missing from the literature.

One conclusion from that policy assessment is that there are policies that, when viewed through the prism of transportation issues, have more promise than urban design. If the goal is simply to reduce traffic congestion or improve air quality, urban design should not be the first place that policy-makers look.

Yet urban design is more than transportation and should be evaluated for its broader, community-building goals. Neighborhood design standards and transportation infrastructure projects, because they are so long-lived, are inseparable from city building more generally. A city's form, and some would say its community spirit, is shaped by its design and transportation infrastructure. This is the basis for the tendency to link the New Urbanism, for example, to transportation planning. Thus, most persons are inclined to confound two somewhat distinct goals—building city forms that will endure and thrive for decades, and managing the more quickly changing transportation problems of today and tomorrow.

The risk inherent in using long-lived urban designs to manage today's congestion and air quality problems is that if situations change, if new solutions become available, or if urban design policies have unanticipated consequences, it is difficult to readjust something as durable as city form. A general maxim for policy is the more flexible the better, and on that count combating transportation problems through city building is about as inflexible as it gets. Thus, we suggest a decoupling, although not a strict separation, of transportation and city building goals.

Possibly the greatest benefits of many of the new urban designs are the more ephemeral goals of livability, public interaction, and community spirit. While those are admittedly difficult to measure, we suggest that too much emphasis on the transportation benefits may sell some designs short.

Certainly urban design and transportation planning are linked, but the difficulty is that we still understand too little about that link to design informed policy. Furthermore, as discussed in chapters 3 and 4, what we do know about the relationship between land use and transportation does not apply well to the small-scale, neighborhood-level design concepts emphasized in most new design strategies.

Our conclusion is not that urban design and transportation behavior are not linked, or that urban design should never be used as transportation policy. Rather, we conclude that we know too little about the transportation impacts of the built environment and that we have other options available that can better meet the transportation planning needs of the immediate future. Yet the link between city building and transportation planning will remain, even if it should be loosely decoupled for policy purposes.

We close by suggesting how future research and scholarship can better tread the terrain between these two sometimes distinct and sometimes related endeavors.

PART II
Travel Behavior

What about cars is bad? In this chapter we consider the reasoning behind efforts to reduce automobile travel, first, by identifying the social costs of car use and, second, by discussing the traditional means for mitigating such costs. Costs are identified either as classic externalities, such as pollution and traffic congestion, or as neighborhood "livability" impacts. Conventional means for managing the former mainly include regulatory or pricing policies; livability issues are addressed by both regulatory policy and other measures. We conclude by discussing where land-use and urban design strategies fit into the conventional wisdom.

2 The Trouble with Traffic

What about cars is bad? In turn, what should transportation planners do?

In the early years of the automobile era, the transportation planner's job was to develop street and highway networks. Sometimes the thinking was as simple as drawing lines on a map to connect concentrations of trip origins and trip destinations, and then building highways along the path that most closely corresponded to those lines.[1] Air quality problems were not conclusively linked to automobile travel until the 1950s. Issues such as the displacement of persons from residential neighborhoods and the impact on habitat were secondary concerns at best until the 1960s. The primary, almost exclusive, focus during the first decades of the automobile era was to build a street and highway network that could accommodate a new mode of transportation.[2]

This began to change by the late 1960s. Planned highway networks neared completion in many cities. At the same time, the broader social costs of transportation became more apparent. Automobile emissions are a major contributor to urban air pollution. Traffic congestion has been a perpetual problem for several decades in most cities. Neighborhoods severed by highway projects often quickly deteriorated. Scholars and policy analysts now ask whether transportation resources are fairly distributed across different segments of society and how transportation access is linked to labor market success. As all of these issues have moved to the fore, transportation planning has increasingly focused on how to manage the social implications of transportation projects.

Modern transportation planning now necessarily focuses as much on managing the social costs of travel as on facilitating travel. Because 87 percent of all trips in the United States in 1990 were by private vehicle (mostly cars and light trucks), the social costs of travel are, first and foremost, the social costs of the automobile.[3] Public concerns regarding air quality, congestion, neighborhood stability, and equity gave rise to new regulatory agencies, technological innovations, and legal frameworks for transportation planning. Yet the demand for cleaner, less congested, more fair transportation systems persists.

This is the context for the new urban designs. They seek, in large part, to address the social costs of automobile travel. Furthermore, because these urban designs strike at what some view as the heart of the problem—an urban form oriented primarily toward the automobile—many view these policies as especially attractive. The goal of the new urban designs is to build cities in ways that manage the social costs of the automobile while enhancing transportation access for persons who cannot or choose not to travel by car. Air quality improvements, congestion reduction, and more livable neighborhoods will, according to this viewpoint, all be achieved in the bargain.

These three benefits—improved air quality, reduced traffic congestion, and more livable communities—constitute the potential link between urban design and transportation. To the extent that land use and design can be used as transportation policy, it is through one or more of those three channels. The air quality and congestion benefits hinge on the idea that urban design can reduce the number or length of automobile trips. The third class of benefits, livable communities, is more multifaceted and requires some explanation.

Proponents of coordinated land-use/transportation policies often cite a broad range of benefits that are not linked to either air quality or congestion (e.g., Katz, 1994). By enhancing pedestrian traffic and public interaction, advocates of the new urban designs hope to create neighborhoods that are lively and diverse in ways that foster a sense of place (Duany and Plater-Zyberk, 1991; Calthorpe, 1993; Katz, 1994; Bernick and Cervero, 1997). Some of this is not related to transportation planning per se. Architectural features and the use of public space are often integral parts of attempts to create communities where residents feel more tied to the neighborhood and thus to each other (Duany and Plater-Zyberk, 1991; Calthorpe, 1993; Katz, 1994). Yet transportation is also important. The increase in community interaction, the use of public spaces that are oriented toward public transit, and the ability to support mixed-use commercial development all rely in part on pedestrian traffic in neighborhoods that conform to the new urban designs (Calthorpe, 1993; Bernick and Cervero, 1997).

To the extent that neighborhood livability is a *transportation* issue, urban design enhances livability by encouraging ostensibly better travel patterns. The sense of place attributed to the new urban designs is, in the eyes of proponents of these plans, enhanced by the pedestrian character and the alternatives to automobile travel that these neighborhoods seek to foster. The architects at the forefront of the new urban design movement have explicitly claimed that much (but not all) of what is desirable in their plans is linked to transportation (Duany and Plater-Zyberk, 1991; Calthorpe, 1993).

The transportation question, then, is, How does urban design influence travel behavior? That is the focus of chapters 3, 4, and 5. The answer, briefly stated, is that we know much less than many think we know. The research task is to develop a framework that furthers our knowledge of the urban design–travel behavior link. That is the necessary first step in evaluating the multifaceted transportation benefits attributed to land use and urban design.

Yet understanding the link between design and travel behavior is not enough. For policy analysis, we care not only whether any one planning strategy can achieve the desired benefits, but also how it compares with other possible alternatives. We must understand the policy context.

This chapter describes the policy landscape against which the new urban designs should be evaluated. For air quality and traffic congestion, that discussion is informed both by a rigorous theory (the microeconomic theory of externalities) and by a rich policy history. For livability benefits, the available theory is less well developed and policy activity is more recent. Yet the benefits of community interaction and sense of place, even if difficult to define and quantify, are important and so must be considered.

We start first with a brief summary of the classic theory of social costs and how it applies to automobile use. Emphasis is placed on air pollution and traffic congestion as examples of the negative impacts of car use that individual drivers often either ignore or undervalue. After that, we discuss neighborhood livability. We then turn to a discussion of the traditional policy instruments that, at least as far as the conventional wisdom goes, seem most effective in reducing social costs to appropriate levels—and how the feasibility of each policy depends on the specifics of the problem at hand. We conclude this chapter by returning to land-use and urban design policies, and where they fit into this framework of problems and solutions.

The Theory of Social Costs

Different sorts of problems suggest different kinds of solutions. Cars and car traffic may be problems, but what kinds of problems? And what can that tell us about which solutions would work?

"Social" costs are conventionally defined as those costs partly or entirely ignored by the persons who cause them.[4] If Bob's trash can falls into the street, spreading trash all over Bob's cul-de-sac, this is a social cost only if Bob doesn't clean up the mess. But if Bob drives in a hazardous manner up and down the street, ignoring the risks to others, then he is imposing what we call a social or an "external" cost. He chose to drive hazardously and either was ignorant of the other costs he imposed on the neighborhood or is consciously ignoring them; in either case, those costs are real and he is neither absorbing them directly nor compensating his neighbors. Because Bob ignores those costs, he likely drives more hazardously than he should.

What if Bob doesn't drive hazardously, but just plain drives? Air pollution from automobiles is a classic case of social cost, often known formally as an externality. In the case of air pollution, the social cost is the harm experienced by persons who must breathe air fouled by the driver's automobile emissions. The key point is that if individuals make driving decisions based only on the benefits and costs that they personally experience, they will not account for how their driving pollutes the air, and will thus drive too much from society's perspective.

Traffic congestion is another example of the external costs of travel, and can be analyzed within this same framework (e.g., Rothenberg, 1970; Heikkila, 1994). Once a highway is congested, each additional driver slows traffic flow further. When balancing the costs and benefits of trips, drivers presumably take into account the speed with which they can travel.[5] Yet each car on a congested road slows travel for all other drivers—an external cost that most will not adequately consider, if at all. So for traffic congestion, like air pollution, private markets will yield more than the socially optimal amount of driving.

Not all social costs of driving are externalities. Neighborhood decay and the displacement of individuals involve the question of how to distribute resources and costs equitably across different groups. Access to employment involves similar concerns. These issues raise important questions of fairness and access to transportation resources, but depending on specifics, external costs may or may not be involved.

Finally, accident costs, estimated by some to be among the largest social costs of driving (Small, 1992, pp. 78–81; Small and Gómez-Ibáñez, 1996, p. 32), may not be external costs to the extent that well-functioning insurance markets provide protection against the risk of loss, harm, or death from those events.

Among all these costs, air quality and traffic congestion are highly visible and are often central both to transportation policy debates and to the discussion about a link between urban design and transportation. So analyzing the benefits of urban design within the context of externalities incorporates an important part of the policy argument.

The externality problem has been well studied by economists, and traditional regulatory solutions can be grouped into three categories: price regulation of externalities, quantity regulation of externalities, and mandated innovations to control externalities. A fourth approach, changing travel behavior, also exists.

Price Regulation of Externalities

The essence of the externality problem is that individuals do not face the full social cost of their actions.[6] In well-functioning markets, the price of a good reflects the resource cost of producing that good. In markets with externalities, the price of a good is below the full resource cost of production because the price does not include the external harm (or social cost).

The heart of the traffic externality problem is this disjunction. People will drive too much when driving is too cheap—too cheap in the precise sense that individuals do not pay for the full social costs of their driving.

Price regulation is an attempt to raise the cost of driving by an amount that, ideally, would exactly reflect the external harm from the air pollution and congestion created by a car trip.

If persons faced the full social cost of their actions, self-interested choices would balance social costs and benefits to achieve a socially optimal quantity. One commonly proposed scheme for doing this is to tax activities that cause externalities. In the cases of air quality and traffic congestion, drivers would ideally pay an "emissions tax" pegged to the harm from the air pollution caused by their driving and "congestion tolls" based on the delay costs they impose on other drivers.

Quantity Regulation of Externalities

While raising the cost of driving will no doubt reduce driving, regulators can also simply mandate that the externality be reduced without changing prices. For example, the 1970 Clean Air Act Amendments set maximum limits for concentrations of harmful pollutants within air basins. Theoretically, this is akin to identifying the optimal quantity of either driving or, in the case of the 1970 Clean Air Act Amendments, the associated externalities, and mandating that the quantities of the externality not exceed the optimal amount. While this can lead to the same outcome as price regulation, quantity regulation typically requires considerably more information and monitoring, themselves both costly, so economists usually recommend price regulation.[7] Yet most environmental regulation in the United States has been quantity regulation, in the spirit of, for example, the Clean Air Act Amendment's maximum limits of concentrations of pollutants in air basins.

Mandated Innovations to Control Externalities

Externality regulation often involves mandated innovations that make the production or consumption of a product cleaner. This has been the most important source of reductions in automobile emissions. Beginning with California's requirement that 1963 model year vehicles have emission controls, several mandated innovations have led to cleaner burning cars.[8] These regulations are technical fixes that do not directly limit the amount of driving. Nor do the regulations have the intent of making driving more expensive, although the cost of complying with these regulations leads to some increase in the cost of vehicle ownership. Instead, this regulatory approach forces the adoption of cleaner technologies that have not yet penetrated the market.

Changing Travel Behavior

In the case of automobile travel, there is a fourth approach that does not fit well into the traditional threefold typology outlined above. Regulations can attempt to change travel behavior, rather than the technology of travel. This might take the form of inducements to carpool, subsidies to public transit, higher parking fees, or more extreme policies such as

Mexico City's no-drive days that ban nearly 20% of that city's cars from the streets on each weekday.[9]

Some urban designs and land-use regulations explicitly attempt to change travel behavior in ways that will lead to air quality improvements and reduced congestion. To the extent that those designs mandate or encourage city building techniques, they are similar in spirit to other technological innovations. In effect, such designs propose to achieve air quality improvements by tinkering with land-use patterns, analogous to the way engineers have achieved air quality improvements by changing automobile exhaust systems.

Alternatively, to the extent that urban designs change the price of travel, they share characteristics of price regulations. Yet the built environment tends to adapt much more slowly than do prices (which can be changed quickly by government tax and subsidy policy) and often more slowly than does vehicle technology. Given this time lag, and the potential difficulty in changing urban form, it is important to assess how the new urban designs compare to other available policies that can regulate both air quality and traffic congestion externalities.

Air Quality Regulation and Automobile Emissions

The first widely publicized link between air quality and automobile emissions was southern California's experience with smog. In the early 1940s, Los Angeles first experienced severe spells of air pollution, then called "gas attacks," which cut visibility to a few blocks. Residents soon noticed that the pollution, popularly called "smog," also irritated the eyes and created other discomforts. Intense air pollution problems were first attributed to industrial sources, and those sources were the focus of public ire and government regulation in southern California (South Coast Air Quality Management District, 1997, p. 2).

By the late 1940s, it was clear that industrial sources were not the only contributor to Los Angeles's smog, but the role of automobile exhaust remained unclear. In 1952, Professor Arie J. Haagen-Smit of the California Institute of Technology demonstrated that hydrocarbons and nitrogen oxides, both components of automobile exhaust, react in the lower atmosphere to form ozone, an important component of Los Angeles's smog (South Coast Air Quality Management District, 1997, pp. 6–11). That work was initially controversial. In a March 3, 1953, letter to Los Angeles County Supervisor Kenneth Hahn, the Ford Motor Company stated that automobile exhaust vapors "are dissipated in the atmosphere quickly and do not present an air pollution problem" (Hahn, 1967, p. 4). Additional evidence soon turned the tide of both scientific and public opinion, and the automobile assumed a prominent role in southern California's smog control efforts (South Coast Air Quality Management District, 1997). In 1961, California required exhaust control equipment on new automobiles sold in the state beginning in the 1963 model year. Like many of the regulatory innovations to follow, California's action preceded national emission control regulations.

At the national level, the 1970 Clean Air Act Amendments gave the Environmental Protection Agency (EPA) authority to develop and enforce national ambient air quality standards (NAAQS). The NAAQS are maximum allowable levels of six different atmospheric pollutants. Each pollutant is measured within a geographic region, or air basin, which typically conforms to a metropolitan area. Among NAAQS pollutants, automobiles account for roughly 60% of all carbon monoxide (CO) emissions in regulated air basins and are an important source of nitrogen oxides (NO_x) and volatile organic compounds (VOCs, the most important of which, for air quality purposes, are hydrocarbons) (U.S. EPA, 1996). Nitrogen oxides and volatile organic compounds react to form ozone in the lower atmosphere.

While the EPA develops the standards for pollutants and monitors air quality under the NAAQS, states have the responsibility for bringing air basins into compliance with the regulations. States must develop state implementation plans (SIPs), frameworks for bringing their air basins into compliance with the NAAQS. Most large urban areas are out of compliance for one or more NAAQS pollutants. For example, 108 counties totaling 70 million residents exceeded the NAAQS levels for ozone in 1995. Within those counties, the EPA has classified 22 air basins as extreme, severe, or serious ozone noncompliance regions (U.S. EPA, 1996, p. 39). When nonattainment for at least one atmospheric pollutant is considered, there are approximately 127 million persons nationwide living in EPA-classified nonattainment areas (U.S. EPA, 1996, p. 60).

In the United States, the penalty for being out of compliance is potentially severe. The 1990 Clean Air Act Amendments specify sanctions for states that contain regions not in compliance with EPA standards. Those sanctions include the possibility of losing all federal highway funds except those grants specifically related to safety and air quality goals (Shrouds, 1992, pp. 27–29; Erbes, 1996, p. 3). Because highway funds are a large source of federal grants for lower levels of government, the prospect of losing that money looms large over state and local transportation decisions.[10]

The 1970 Clear Air Act Amendments also gave the EPA the authority to regulate automobile tailpipe emissions. The EPA mandated that automobile emissions of CO, NO_x, and hydrocarbons (or VOCs) be reduced to 10% of 1970 levels by 1975 (Lave and Omenn, 1981, pp. 30–31). This requirement was pushed back several times. These target tailpipe emission improvements were the impetus for several technological improvements in vehicle exhaust systems (e.g., catalytic converters), fuels (e.g., reformulated gasoline that reduces the level of VOC emissions and oxygenated gasoline that reduces CO emissions), and vehicle inspection programs. These have generally been successful in producing cleaner burning cars.[11] For example, prior to 1968, the average car sold in the United States emitted 84.0 grams of CO per mile and 10.6 grams of VOCs per mile. In 1993, United States emissions control standards required that new vehicles emit no more than 3.4 grams of CO per mile and 0.41 grams of VOCs per mile—a 96% reduction in the tailpipe emissions of those two pollutants (Small and Kazimi, 1995, p. 10).

Air quality has improved in most United States metropolitan areas during the last two decades. The EPA calculates a pollutant standards index (PSI), based on concentrations of CO, nitrogen dioxide (NO_2), ozone (O_3), particulate matter with a diameter less than 10 micrometers (PM-10), and sulfur dioxide (SO_2).[12] For each of those five pollutants, measured concentrations for metropolitan areas are converted into a scale that ranges from 0 to 500. Higher index values indicate higher atmospheric concentrations, and 100 corresponds to the NAAQS standard. Atmospheric concentrations are measured daily, and the pollutant with the highest value for each day is reported as the PSI for that day.

For all metropolitan areas with populations exceeding 200,000, the PSI index exceeded 100 on a total of 1,584 days in 1986; however, the total number of days the PSI exceeded this level dropped to only 707 by 1995 (U.S. EPA, 1996, pp. 64–65). Moreover, the maximum concentration of NAAQS pollutants (measured on a daily basis) either dropped or did not change (within accepted ranges of statistical significance) in virtually all urban areas in the United States during this period.[13]

The Los Angeles–Long Beach metropolitan area, often considered the nation's smog capital, showed statistically significant decreases for all six NAAQS pollutants from 1986 through 1995 (U.S. EPA, 1996, p. 146). During the past two decades, the number of smog alert days in Los Angeles has dropped dramatically. The NAAQS standard for ozone is 0.12 parts per million (ppm), a Stage I smog alert is called when the ozone concentration exceeds 0.20 ppm, and a Stage II alert occurs when the ozone concentration exceeds 0.35 ppm. In 1978, the Los Angeles area had 117 Stage I alerts; in 1988 it had 77 Stage I alerts; in 1993, 23 Stage I alerts; in 1996, 7 Stage I alerts (*Los Angeles Times,* 1992; Cone, 1996).

In 1996, for the first time since records have been kept, Los Angeles did not have the worst smog day in the nation.[14] Stage II alerts, which trigger recommendations that at-risk persons remain indoors, were not unusual in the Los Angeles area in the 1970s. There has not been a Stage II alert day in the Los Angeles metropolitan area since 1986 (*Los Angeles Times,* 1992).

The recent improvements in air quality have been achieved mostly by mandating technological improvements. In terms of transportation, a series of mandated emissions technologies and changes in gasoline formulation led to much cleaner burning cars. Almost all of the reduction in automobile pollutants can be attributed to these technological innovations, rather than changes in driving behavior or pricing policy. In fact, United States citizens drive more than ever before on a per-household basis, and as recently as the late 1990s inflation-adjusted gasoline prices were lower than they have been in over twenty years.[15]

Yet in air quality regulation, the easy gains have been made. In the 1960s, several low-cost emission control technologies were available but had not been adopted. Mandating the adoption of those technologies produced dramatic air quality improvements. Further gains will require tougher choices (e.g., Howitt and Altshuler, 1999). Some persons argue that one of those tough choices will involve changing travel behavior—and recent policy activity has in general increased the attention on travel behavior.

The 1990 Clean Air Act Amendments, the 1991 Intermodal Surface Transportation Efficiency Act (ISTEA), and its renewal as the 1998 Transportation Equity Act for the 21st Century (TEA-21), all give renewed attention to travel behavior. A few local governments charged ahead with ambitious travel behavior modification programs developed in the 1980s. One of the most notable experiments among these was the trip reduction policy of the South Coast Air Quality Management District (SCAQMD), in the greater Los Angeles area.

In California, air quality management district boundaries generally conform to air basins, and the districts have authority to develop and implement plans to bring their air basins into compliance with state and federal clean air regulations. In the late 1980s, the SCAQMD developed a trip reduction program known as Regulation XV. This regulation required firms with more than 100 employees to file trip reduction plans to support carpooling, alternatives to automobile travel, and other incentives that would encourage their employees to commute by means other than single-occupant vehicles. The program was controversial, often because the regulatory burden was viewed by many firms as unduly large, and compliance with trip reduction plans is now voluntary.[16]

Despite the controversy and the deemphasis of Regulation XV, that program, similar local experiments elsewhere, and the continued and stricter federal requirements for cleaner air increased the focus on travel behavior in relation to air quality problems.

Traffic Congestion

Traffic congestion is a common problem in most large urban areas. Lindley (1987) estimated that congestion costs, in terms of lost time and extra fuel consumption, totaled $9.2 billion in the United States in 1984. Estimated congestion costs typically exceed estimated pollution costs from automobile emissions, making congestion and accidents the most costly external effects of automobile travel (Small, 1992, p. 84; Small and Gómez-Ibáñez, 1996).[17] Public opinion polls show that traffic congestion is often cited as one of the most pressing policy problems in many urban areas.[18]

The Texas Transportation Institute (TTI) has measured traffic congestion in U.S. urbanized areas since 1982 (Schrank and Lomax, 1997). Boarnet, Kim, and Parkany (1998) developed a congestion index for counties and urbanized areas in California for the years 1976 through 1994. Both indices show a general worsening of congestion over time. This is consistent with public perceptions that congestion levels are growing increasingly worse.

Yet Gordon, Richardson, and Liao (1997) note that the rank order correlation between TTI's roadway congestion index and self-reported work-trip travel speeds (from the Nationwide Personal Transportation Survey) in a sample of large U.S. metropolitan areas is only 0.09. In related work, Gordon, Richardson, and Jun (1991) show that automobile commute trip duration either fell or was unchanged from 1980 to 1985 in all of the 20 largest U.S. metropolitan areas. Gordon and Richardson (1993) report that average commute times have remained roughly constant (at around 25 minutes) in Los Angeles from 1967 through 1990 (the last year for which data were available at the time of their report).[19] Overall, while both common perception and congestion indices suggest that congestion is growing worse, commute times have hardly changed for several years in many U.S. urban areas.

The explanation offered by Gordon, Richardson, and Jun (1991) for this seeming paradox is that persons and firms have relocated from central cities to less congested urban areas. Thus, while network congestion worsens, more travel now occurs on the relatively less congested urban fringe. They contend that the two phenomena are causally related—that firms and residents relocate to suburban areas in part to avoid central city traffic congestion.[20] Congestion is in part its own solution, in this view, as persons and firms move away from congested areas.

Does urban decentralization eliminate the need for a policy response to congestion? Congestion is an externality and moving to avoid it entails additional costs. According to Small and Gómez-Ibáñez (1996), it is unlikely that urban relocation decisions, which themselves entail moving costs, would represent an optimal policy response to congestion. Rather than suggesting that policy-makers not worry about traffic congestion, the value in the analysis of Gordon, Richardson, and colleagues is in pointing out the need for a balanced yet serious approach to congestion problems, rather than reacting to "doomsday scenarios."[21]

Policy responses to traffic congestion can be grouped into two classes —those increasing traffic capacity, that is, the supply of highways and streets, and those reducing travel demand on congested arteries at peak hours. Recent experience in the United States suggests that supply expansion will not be an option in the near future except in isolated instances. After large sums of money were invested in highway construction during the 1950s and 1960s, the value of the nation's stock of highways has remained roughly constant for the past two decades (Gramlich, 1994, p. 1179). One explanation is that highway building is increasingly subject to a fiscal squeeze. Highway construction has grown more expensive, not the least reason being that rising land prices have made right-of-way acquisition prohibitively costly in many urban areas. At the same time, gasoline tax revenues, the primary source of highway construction funds in the United States, have not kept pace with either vehicle miles traveled or inflation. The result is that fiscal realities do not favor a return to the large highway construction programs of past years (Taylor, 1995).

Even if funding for more highway miles were available, conventional wisdom holds that urban areas cannot build their way out of congestion problems. Downs (1962) famously stated that travel demand on unpriced congested freeways rises to meet capacity, an idea since called Downs' Law. In this view, there is latent (or unrealized) demand for travel on congested highways. If new lanes are added, congestion problems might be lessened in the short run. But that reduced congestion will attract drivers who previously used other routes, traveled at different times of the day, or used other modes. Soon this latent demand will lead to congested conditions on the improved highway (Small, Winston, and Evans, 1989; Downs, 1992). Recent studies have found empirical support for the idea that highway supply expansions induce additional travel demand (Hansen and Huang, 1997; Noland, 1998).

Given the problem of latent demand, and the inherent difficulty in solving congestion problems through highway construction, transportation planners began to focus on transportation demand management (TDM). This includes policies such as the SCAQMD's Regulation XV, which explicitly seeks to reduce work trips. Other policies, such as carpool lanes, are hybrids; they increase supply (since carpool lanes are often new capacity), but they also seek to change travel behavior by encouraging people to share cars. Unfortunately, the evidence suggests that dedicated lanes for car-poolers deliver only small gains in congestion reduction (Giuliano, Levine, and Teal, 1990). Similarly small driving reductions have been credited to the early years of the SCAQMD Regulation XV program.[22]

Given the difficulties with solving congestion by increasing highway supply, and the generally small impact of traditional TDM policies, pricing solutions are lately receiving more attention. In theory, if correctly calculated, corrective congestion tolls provide an economically efficient solution, as individual drivers will drive only when the value of their trip exceeds both their private costs (time and money costs) and the delay cost that their driving (during congested time periods) imposes on others (e.g., Mohring and Harwitz, 1962; Vickrey, 1963; Small, Winston, and Evans, 1989).

Congestion pricing has several potential advantages. Because it manages travel demand, it is both less costly and more likely to succeed than are supply expansions that might later be swamped by latent demand. Congestion costs can readily be calculated (e.g., Keeler and Small, 1977), and once the proper tolls have been charged, congestion levels will drop to a socially optimal amount. With electronic toll collection technology, it is now possible to detect passing vehicles and deduct the proper toll from a prepaid or credit account (Sullivan and El Harake, 1998). Because congestion pricing can lead to more efficient use of the existing highway infrastructure, there is some evidence that it can produce modest gains in economic output and productivity in urban areas as well (Boarnet, 1997b). Last but not least, because the price can be adjusted to manage changing levels of congestion, pricing is one of the few policies that can potentially provide more than short-term congestion relief.

Despite these advantages, congestion pricing had for years been a political dead-end in the United States and most other countries. Proposals to charge for previously free travel continue to meet with vehement opposition, and there remain only a handful of congestion pricing projects in the world.[23] The conventional wisdom holds that the benefits of congestion pricing are abstract efficiency gains, which are long term and diffused across many persons. The costs, in terms of paying for previously unpriced travel, are immediate and obvious to drivers. Thus, congestion pricing proposals have generated strong opposition and often little support beyond academic circles (Giuliano, 1992; Wachs, 1994).

Yet the times might be changing.

After having met with stiff political opposition for years, congestion pricing has been implemented on the HOV (or carpool) lanes opened in the median of State Route 91 (SR-91) in southern California in 1995. The SR-91 cuts through a canyon connecting growing bedroom communities near Riverside with the employment centers of Orange County, south of Los Angeles. Because of geographic constraints, there are few alternatives to traveling the SR-91, and at peak hour congestion delays have been notoriously long (Sullivan and El Harake, 1998).

The SR-91 project is unique in several ways. The franchise for the HOV lanes was granted to a private contractor as part of four public-private demonstration projects authorized by legislation passed by the State of California in 1989 (U.S. Department of Transportation, 1992; Gómez-Ibáñez and Meyer, 1993). The California Private Transportation Company (CPTC) built and now operates the two new lanes in the median of the preexisting freeway. The CPTC is financing their project through toll revenues.[24] Early during the project planning, the CPTC decided to charge higher tolls during peak hours, both to collect more revenue and to manage congestion on the tolled lanes. As of mid-1999, the tolls varied from $0.75 to $3.35 for the ten-mile stretch based on time of day.

This application of congestion pricing never met with much serious opposition.[25] Unlike other proposed projects, the tolled lanes on the SR-91 provided new capacity immediately adjacent to existing free lanes. Travel time in the free lanes dropped by as much as twenty minutes for that 10-mile segment after the tolled lanes opened, as some traffic diverted to the new capacity (Ortner, 1996; for related discussions, see Mastako, Rilett, and Sullivan, 1998; Sullivan and El Harake, 1998). The SR-91 project demonstrates two important points: (1) In a highly congested corridor, persons will pay to reduce their travel time, and (2) if the tolled lanes divert traffic from existing free lanes, even those who do not wish to pay are made better off by the toll facility (Sullivan and El Harake, 1998).

The public acceptance of congestion pricing in the SR-91 corridor has led to much discussion of other ways to use tolls to manage travel demand. One of the more promising ideas is high occupancy toll (HOT) lanes, such as those implemented in the northern suburbs of San Diego, California. The HOV lanes on the Interstate 15 north of San Diego had been underused for several years prior to the implementation of a HOT lane experiment. Local officials decided to use the excess capacity by allowing single-occupancy vehicles to "buy into" the carpool lanes, initially with the purchase of a monthly pass (Pund, 1997). The revenues from this project are used, in part, to finance express bus service along the corridor, thus alleviating some of the concern that pricing projects inherently favor upper income individuals who can afford to pay for faster travel.[26] Elsewhere in California, proposals for HOT lanes are being discussed in Los Angeles and Orange counties (Stone, 1996; Parrish, 1997).

Overall, the recent success of congestion pricing experiments suggests some scope for broader implementation of that idea (Boarnet, 1999). Whether or not pricing will become an important policy tool in many urban areas remains to be seen. Yet for now the old assumption that congestion pricing is nothing more than an ivory tower fantasy seems unduly pessimistic.

Neighborhood Quality of Life

The link between transportation and a broad range of neighborhood characteristics is an increasingly important area of policy focus. At the federal level, the Federal Transit Administration's (FTA) "livable communities" initiative emphasizes coordinated development near rail transit stations. The goals are several, and include providing transportation options in otherwise automobile-dependent urban areas, increasing the architectural diversity in suburban regions, focusing civic interaction in mixed-use neighborhoods with public spaces, and providing a focus for affordable housing that is tied to alternative modes of transportation. These same goals are cited by proponents of the new urban designs, regardless of whether those designs are tied to transit (Duany and Plater-Zyberk, 1991; Calthorpe, 1993; Katz, 1994).

Still, unlike efforts to address air quality and traffic congestion, the policy activity in this area is almost exclusively local. The FTA's initiative primarily supports and focuses local efforts. Some states passed growth control laws that encourage or require localities to pursue higher density development (see, e.g., Bollens, 1992), but the policy activity in this arena that has been consistent with the new urban designs has typically originated at the local level. California passed a Transit Village Development Act (Knack, 1995), but the law brings few clear requirements, and again largely encourages and facilitates local planning near rail transit stations.

Furthermore, neighborhood quality of life is inherently a local concept, focusing as it does on local involvement, community control, and a "sense of place" at a scale often far smaller than a city. Given the difficulties inherent in changing existing built environments and political structures, many of the quality-of-life initiatives linked to urban design were pioneered in the context of new, master-planned communities.

Master planning has emphasized social goals and "livability" for decades, but the idea of a livable community has changed. The Levittowns of the 1940s and 1950s were designed to respond to the large postwar demand for suburban housing. Those early tract developments were sometimes viewed as providing a respite from the more crowded conditions in cities. In the 1960s, when race relations came to the fore, Columbia, Maryland, was designed to be a racially integrated community. In the 1990s, public concern evolved to include a focus on recapturing some of the neighborliness and community spirit associated with small towns. The desire to build communities rather than tract houses is reflected in the plans for Laguna West, outside of Sacramento, California, Otay Ranch near San Diego, and Celebration, adjacent to Orlando, Florida.

All these planned communities intended to achieve social goals in part via urban design. Certainly, one's quality of life is influenced by the environment in which one lives, but human behavior is more complex than simply a reflection of the neighborhood, and attempts to engineer social change through neighborhood building must acknowledge that.[27] The important point is to evaluate carefully the way in which neighborhood characteristics improve individual lives. For transportation, that careful evaluation of neighborhood quality of life and urban design must hinge on travel behavior. To the extent that *transportation-related* benefits flow from urban designs, it is because those designs alter the way people travel.

Summary

The primary transportation benefits associated with reducing car travel are threefold—air quality improvements, reduced traffic congestion, and improved neighborhood quality of life. The first two are external costs associated with automobile travel, and the benefits of land-use planning and other urban design strategies can be evaluated within the theoretical framework of externality regulation. More important, both air quality and traffic congestion were the focus of considerable policy activity the past several decades. Urban design and other land use solutions to those problems should be evaluated within the context of a range of policy alternatives, and their track records, some of which have delivered dramatic successes in the past.

But even if they make sense as either direct or indirect schemes for reducing the social costs of car travel, what can land-use planning and urban design actually accomplish? Chapters 3–5 review what we know about how urban form influences travel.

If reducing automobile travel has social benefits in some settings, how and when can urban design help? In this chapter we review recent studies addressing this question. We begin with a summary of urban spatial theory and other conceptual frameworks explicitly linking urban structure to travel. We then look at studies examining these theories, some by comparing alternative scenarios and others by analyzing observed behavior to formally test behavioral hypotheses.

3 Studies of Urban Form and Travel

Does the built environment affect how often and how far people drive or walk or when they will take the bus or the train? If so, how?

A lively, expanding literature continues to investigate the potential for causal links between urban design and travel behavior, yet there remain many gaps and considerable disagreement. Our purpose here is mainly to identify what past research has to say on these questions. We also try to explain why these studies reach different conclusions and how and where this work might be usefully improved.

The first, and perhaps best-known, group of studies on this topic investigates how travel behavior and travel investment affect land use. There is also a long if more recent practice of viewing these links from the opposite direction; that is, how does land use influence urban travel? We consider this second question in more detail following a brief review of the first.

The Influence of Transportation on Urban Form

Though not our focus, most questions about land-use/transportation links over the past century concern the influence of transportation infrastructure on development patterns. Analysts ask how highways and mass transit contribute to decentralization trends, how they affect the local balance of jobs and housing, or how they affect the pattern of commercial investment (see, e.g., the reviews in Gómez-Ibáñez, 1985b; Giuliano, 1989, 1991, 1995a, 1995b; Cervero and Landis, 1995).

The basic idea is this: People choose their homes and locate their businesses based in part on their proximity to work, other potential destinations, and the markets for their products and labor generally (see, e.g., Von Thunen, 1826; Weber, 1928; Losch, 1954; Alonso, 1964; Muth, 1969; Mills, 1972; Solow, 1973; Fujita, 1989; Anas, Arnott, and Small, 1997). That is, the cost of transporting people and things over space depends on the distances and resources required. Once these costs are fixed, perhaps by the establishment of a central downtown or transshipment point, the price of land at each location is determined by demand. This in turn is determined, again in part, by how much money one has left after accounting for the transportation costs associated with that location. That simple concept drives many of the core results in classical urban economic theory: Land and housing prices will tend to decline with distance from where people want to go.

Moreover, the more expensive the land at a given location, the more likely a given site will be developed densely as builders trade off construction costs against unit land costs. Thus, densities are also expected to decline with distance from central locations. In equilibrium, households and firms will choose to locate based on their individual evaluation of the market trade-off between transportation costs and their demand for space.

In this view, residential suburbs are the outcome of three primary factors: declining travel costs due to freeway construction or transit networks, rising per capita incomes combined with the choice by some for larger lots at the expense of higher commuting costs, and the inability of the poor or those facing housing market discrimination to follow (Mieszkowski and Mills, 1993). This view is based on a model of urban areas with a dominant employment center surrounded by residential suburbs. Yet employers often choose to locate outside downtown, either for cheap land or to reduce the commutes of their increasingly suburban workforce. Many suburbs are now job centers that rival and sometimes surpass the employment levels and economic importance of downtown.[1] Thus, standard urban location theory no longer applies in its cleanest form to many metropolitan areas.

Most cities have many centers and subcenters, pulling the conventional single-centered urban structure into a variety of multinucleated forms (Giuliano, 1989, 1991, 1995b; Gordon, Kumar, and Richardson, 1989a; Giuliano and Small, 1991; Boarnet, 1994a, 1994b; Anas, Arnott, and Small, 1997). These forces are accentuated by the growing incidence of multiworker households and the relatively unpredictable location of future jobs when job locations are moving among subcenters in a common labor market (e.g., White, 1988; Zax, 1991, 1994; Zax and Kain, 1991; Giuliano and Small, 1993; Rosenbloom, 1993; Crane, 1996c; Gordon and Richardson, 1996; Van Ommeren, Rietveld, and Nijkamp, 1997).

But that is not even half the story. In addition to the effects of changing employment locations and ongoing decentralization of workers on urban form, it is increasingly evident that the journey to work is no longer the defining travel experience. Considerably fewer than half the automobile trips today are commute trips, and the emerging changes in commute behavior described above are unlikely to have any substantial effect on urban form, most of which is well established, except at the periphery and for infill (Giuliano, 1989, 1991, 1995; Rosenbloom, 1993; Dunphy, Brett, Rosenbloom, and Bald, 1997).

Initial investments in transportation infrastructure, such as the first interstates, almost certainly led to significant local development impacts (Giuliano, 1995a,b; Boarnet, 1997a). The effects of subsequent investments are less clear. For example, a recent study of the impacts of the San Francisco area BART commuter rail system concluded that, some twenty years following its introduction, line and station locations did impact the development of the communities in which they were located (Cervero, 1995, 1994b; Cervero and Landis, 1997). But that evidence suggests that the main form of the impact was to anchor and guide economic activity rather than to generate it. It is difficult to establish that a particular rail station did anything more than prepare a site for the economic development that would have taken place in that city anyway. Other studies are even less conclusive, particularly in well-developed urban areas (Cervero, 1989a; Moon, 1990; Moore and Thorsnes, 1994; Giuliano, 1995b; Bernick and Cervero, 1997; Bollinger and Ihlanfeldt, 1997).

Hence, the influence of transportation investments on urban form likely varies with its timing, among other things. The linkage and even causality are clear in some cases and at some times, but there is also strong evidence that such effects are diminishing and perhaps inconsequential in many instances.

The Influence of Urban Form on Travel

What if we turn the causality around? Engineers and planners have long employed, with much confidence, estimates of trip generation rates and other travel behaviors associated with alternative development patterns (e.g., Olmsted, 1924; Mitchell and Rapkin, 1954). This practice continues, with refinements to improve the reliability and flexibility of such standards (e.g., Institute of Transportation Engineers, 1991, 1997). That is, the people who actually build our streets and cities assume, as a matter of course, that the built environment does indeed influence travel behavior.

Where the research examined below departs from the simple calculation and application of engineering standards is primarily in its preoccupation with the travel impacts of alternative residential patterns, and its attention to other measures of travel behavior beyond trip generation and parking requirements. Rather than merely estimate that an average two-bedroom apartment generates *X* fewer car trips per day than does a three-bedroom house, the recent literature is more aware of how this estimate might vary depending on circumstances. In particular, it focuses on land-use factors, for example, population density, employment location, mixed land uses in the neighborhood and region, and the local street configuration. In addition, these factors are associated with outcome measures that include vehicle miles traveled (VMT), car ownership rates, and mode choice.

While the goal of past research was mainly to predict travel flows for given land-use patterns, the goal of the more recent literature is to understand how travel behavior might be influenced by manipulating urban form. The focus has subtly but importantly shifted from *prediction* to *prescription.*

The motivating question now, implicitly and often explicitly, is how to design neighborhoods and the larger community to reduce automobile use. The intent, as discussed in chapters 1 and 2, is to stimulate the interaction of residents by increasing pedestrian traffic and generally improving neighborhood charm, as well as to reduce air pollution and traffic congestion. That goal has given rise to a large but still quite new body of studies on whether and how changes in land use and urban design can cause changes in travel behavior.

In organizing a summary of any literature it would be useful to propose a typology, but there is no one best rationale for doing so in this instance. These studies can be usefully organized in any number of ways, for example, by travel purpose (journey-to-work travel vs. shopping vs. trip chains, etc.), by analytical method (simulations vs. regressions, etc.), by the characterization and measures of urban form (trip ease vs. street layout vs. composite measures of density, accessibility, or pedestrian features, etc.), by the choice of other explanatory variables (travel costs vs. travel opportunities vs. characteristics of the built environment or of travelers, etc.), or by the nature and level of detail in the data. All are effective schemes for distinguishing among various strategies for identifying and measuring the influence of land use and urban design on travel. And each offers different insights into how and why different approaches yield different results.

Table 3.1 lists these options, divided into four categories. Most attention, historically, has been with the first two columns as effect and cause, respectively. The first lists the travel behaviors under examination, as measured in the literature. They include total travel, trip generation rates, car ownership, mode choice, and the length of the journey to work, among others. The second column lists the urban form and land-use measures that might influence travel behaviors. They feature measures of population and employment density, the land-use mix, the street pattern, and the balance of jobs and housing. Studies commonly attempt to identify and verify the linkages between these two columns and their parts.

The third column lists the most common means used to study these questions: simulations, descriptions, and multivariate statistical analysis. How these differ is discussed in more detail below, but table 3 summarizes a few points. Simulations are based either on entirely hypothetical situations, and thus succeed or fail depending on the validity of their assumptions, or on more complex combinations of assumed and forecast behaviors. These are useful and interesting exercises, but there are certain questions they are ill-equipped to address. For example, they cannot test hypotheses regarding the effect of land use on travel behavior. On the other hand, simulations do illustrate how alternative scenarios

Table 3.1: A Listing of Outcomes, Questions, and Methods in Studies of Urban Form and Travel

Travel Outcome Measures	Urban-Form and Land-Use Measures	Methods of Analysis	Other Distinctions and Issues
1. Total miles traveled (e.g., VMT) 2. Number of trips 3. Car ownership 4. Mode (car, rail, etc.) 5. Congestion 6. Commute length 7. Other commute measures (e.g., speed, time) 8. Differences by purpose (e.g., work vs. nonwork, regional vs. local)	1. Density (e.g., simple residential/ employment, or more complex "accessibility," etc., measures) 2. Extent of land use mixing 3. Traffic calming 4. Street and circulation pattern 5. Jobs-housing and/or land-use balance 6. Pedestrian features (e.g., sidewalks, perceived safety, etc.)	1. Simulation (i.e., simple hypothetical impacts based on assumed behavior, or more complex integrated land-use/traffic impact models based on forecasts) 2. Description of observed travel behavior in different settings (e.g., commute length by city size) 3. Multivariate statistical analysis of observed behavior (i.e., ad hoc correlation analysis or model specified and estimated according to behavioral theory)	1. Land use and urban design trip origin vs. trip destination vs. entire route 2. Composition of trip chains and tours (e.g., errands on commute home) 3. Use of aggregate versus subject-specific data

VMT = vehicle miles traveled.

compare given certain behavioral assumptions. For that reason, they are used extensively for transportation investment alternatives analysis.

Descriptive studies provide hard data on real behaviors in different situations. For example, how do people who live downtown get to work and how does this compare with the commute mode choice of suburban residents? Their purpose, and strength, is in showing us what is happening at a particular place at a particular time. Unfortunately, this approach rarely tells us much about why people behave as they do, particularly regarding an activity as complex as travel.

Another class of methods includes multivariate techniques, usually some form of regression analysis. These are very useful for travel studies since so many factors are at play. Where people want to go and how they plan to get there depend on their resources, the transportation network in place, their access to a car, bus, or commuter rail system, the needs, demands, and desires of their families, their demand for the goods that travel can access, gasoline prices, bus fares, and so on. Lots of things appear to matter, and multivariate methods are well suited to analyzing such situations.

It is useful to distinguish between two kinds of multivariate models, as the third column of table 3.1 indicates. In the first model, one or more of the travel outcomes in column 1 are associated with various land-use and urban-form measures in column 2, perhaps along with other variables believed to help explain travel. A common approach is to regress commute length on a measure of residential density and the demographic characteristics of travelers, and then examine the significance, sign, and magnitude of the estimated coefficient on density (good examples are Frank and Pivo, 1995; Levinson and Kumar, 1997; Sun, Wilmot, and Kasturi, 1998). If the coefficient is significantly negative, the analyst might conclude that commutes are shorter in relatively dense settings, and indeed that perhaps increased development densities would in turn reduce VMT among workers. The great number of studies of this kind have led prominent reviewers to conclude that "every shred of evidence" or "a preponderance of evidence" supports the conclusion that higher densities reduce VMT (Ewing, 1997a; Burchell et al., 1998).

As Crane (1996a, 1996b), Dunphy and Fisher (1996), Handy (1996b), Myers and Kitsuse (1999), and others have pointed out, however, this approach is inadequate in several respects. For one, density is more than a simple feature of the built environment that can be either readily described or easily replicated. It has many significant dimensions, likely too many to capture meaningfully in one or two indices. For another, the explanation for density is itself an important yet often neglected part of the story. VMT may be low in areas of high density for a particular data set mainly because incomes are low in those areas, or because other differences among places that are correlated with density are absent from the data and hence the analysis.

Finally, there is little behavioral content in these analyses to clarify how or why travelers, and potential travelers, select among the set of feasible travel choices. What is generalizable about the factors in one environment that generate more and longer car trips, and in another fewer or shorter trips? While some such studies do attempt to control for different trip purposes (e.g., shopping vs. commuting), trip lengths (neighborhood vs. regional), and demographic variables likely associated with trip demand (income, age, etc.), the approach is typically ad hoc. It has no strong conceptual framework to frame statistical results or systematically make the case for causality outside the data, making both supportive and contrary empirical results difficult to compare or interpret.

An alternative approach to multivariate analysis of these questions would incorporate urban form measures into a transparent behavioral framework that systematically explains travel behaviors. Work of this sort continues to be rare. There is an extensive literature on behavioral choice in travel, to be sure, but it has neglected the role of land use and urban design (e.g., Domencich and McFadden, 1975; Small, 1992; Gärling, Laitila, and Westin, 1998). Some representative studies that do examine the influence of urban form on travel in a consistent behavioral framework are discussed below. Chapters 4 and 5 then build on that work.

Several important studies are not reviewed here, and others are mentioned only briefly, for lack of space.[2] How were the others selected? In some cases, early studies provide an interesting base and context for the state of the literature. In other instances an article may have a particularly provocative result, unique data set, or methodological wrinkle that fits the order and rhythm of the discussion. Overall, the idea is to present a clear picture of what the literature does and what it has accomplished, with citations the reader can investigate further, rather than to recognize the role of each individual scholar, paper, or significant result. Unfortunately, that means little of the hard work and progress reflected in this research receives the attention it deserves in this chapter.

Hypothetical Studies

The world is a very complicated place. It is rarely easy to sort out cause and effect or even what exactly is happening at any point in time, let alone why. The general idea in hypothetical studies is to construct situations, in a strategically simplified yet tightly controlled environment, where different land-use patterns and other urban design features can be linked clearly to travel. The exercise is only artificial to the extent it is incorrect. Say, for example, we simulate a city where 80 percent of the population drives to the grocery story and the remainder walk or take the bus. What happens if we increase the cost of gasoline or parking, or reduce bus fares, or change the subdivision layouts or residential densities so that grocery stores are closer to residents' homes? Several studies have done just that and reported the results as examples of what might happen in real communities that did the same.

As Handy (1996b) points out, hypothetical studies are not intended to explain behavior. Rather, they make certain assumptions regarding behavior and then apply those to alternative situations to see what happens. In general, the results of hypothetical studies applied to the urban design/travel question are unsatisfactory for just that reason. Most existing simulations ignore certain pivotal characteristics of the built environment and of travelers, in our view, and poorly account for feedback, that is, the manner in which travelers respond to changes in their circumstances.

For example, Calthorpe's (1993) assertions about the transportation benefits of his suburban designs depend heavily on a simulation by Kulash, Anglin, and Marks (1990), who found that "traditional" circulation patterns reduce VMT by 57 percent as compared to more conventional networks (figures 3.1–3.3). The usefulness of this result is limited, however, because Kulash and colleagues assume that trip frequencies are fixed. They also assume that average travel speeds are slower in a grid-based network, which in turn requires nonstandard street design standards.

The more elaborate simulation studies of McNally and Ryan (1993),

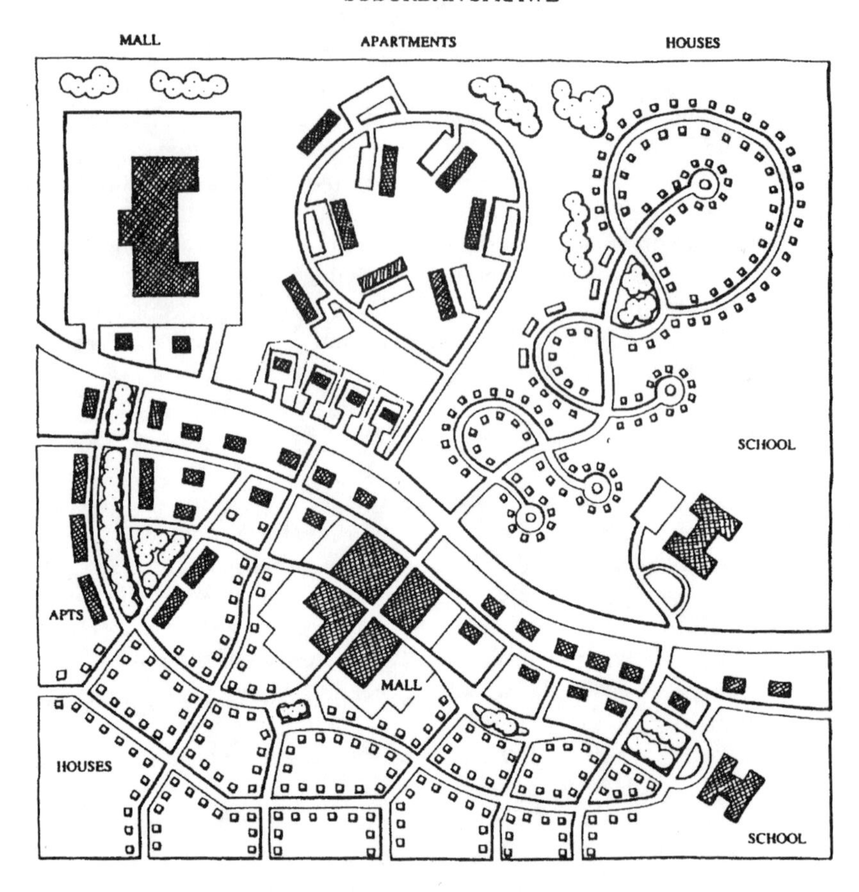

Figure 3.1. A comparison of "suburban sprawl" and "traditional" neighborhood development (Duany and Plater-Zyberk, 1992).

Rabiega and Howe (1994), and Stone, Foster, and Johnson (1992) also tend to focus on whether a more grid-like street pattern reduces VMT.[3] They model the new plans as essentially moving trip origins and destinations closer together, but most hold the number of trips fixed. (Stone, Foster, and Johnson [1992] let trip generation rates change based on assumed differences in the land-use mix in each scenario, and then apply fixed trip rates for each use based on published engineering standards.)

Thus, the studies essentially ask, If a trip becomes shorter, will people drive as far? It is easy to see that the answer is no, but what we learn from the exercise about the expected impact of these schemes is unclear. The result follows directly from the statement of the problem. The simplest example is that as you move average trip origins and destinations closer

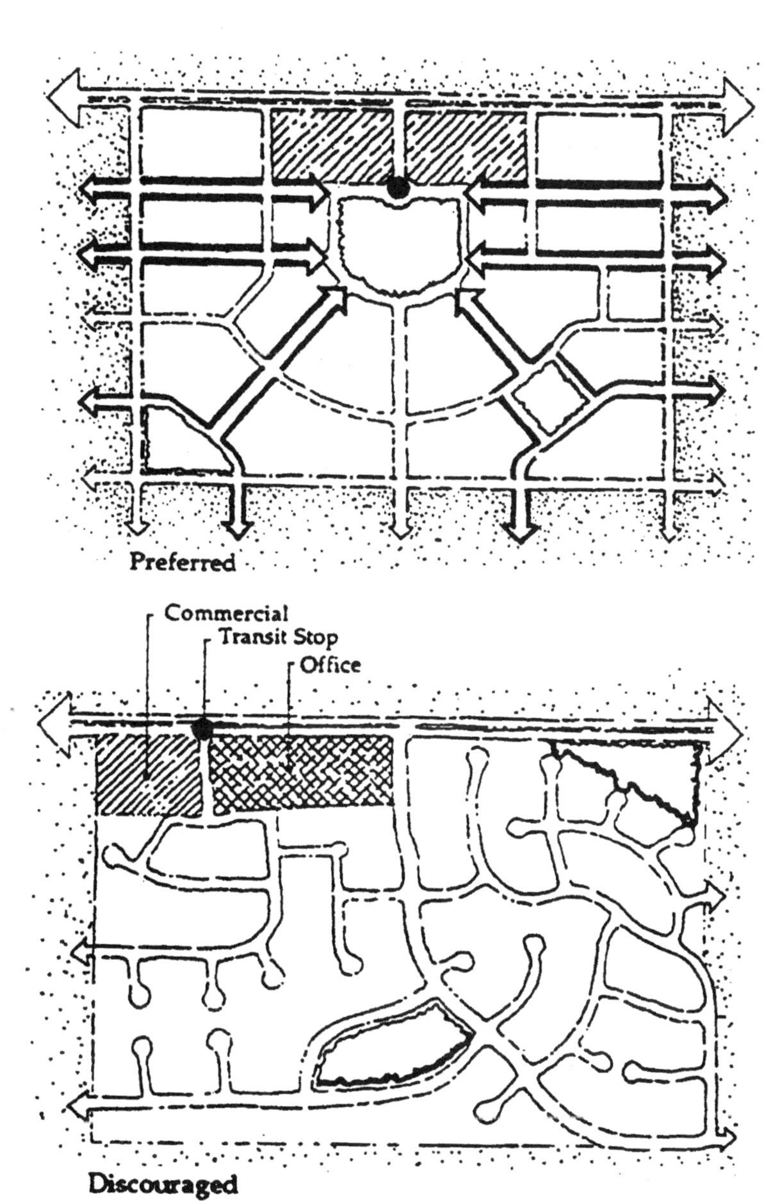

Figure 3.2. A comparison of "preferred" and "discouraged" street and circulation patterns in the "transit-oriented" development guidelines prepared for the City of San Diego by Calthorpe Associates (City of San Diego 1992).

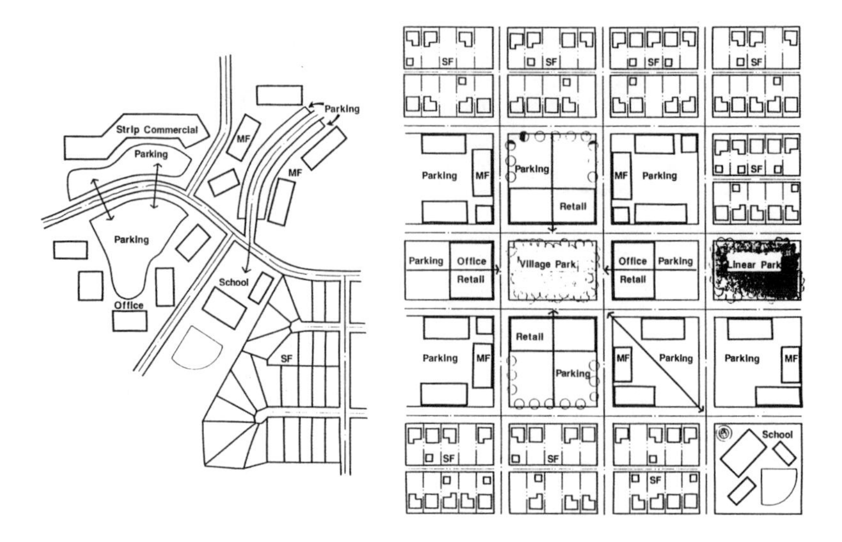

A. Conventional Suburban Development (CSD)

B. Traditional Neighborhood Development (TND)

Figure 3.3. A comparison by Kulash, Anglin, and Marks (1990) of "conventional" suburban development and "traditional" neighborhood development.

together, which higher densities, mixed land uses, and a grid street layout do, trip lengths must decrease on average. The unanswered questions are whether the number of trips and travel mode, or other decisions, are also affected by a change in trip length. These studies typically assume away such responses—apart from what engineering standards imply—though behavioral feedback may be key to understanding what will happen to travel in practice. The lack of behavioral content, a problem shared by virtually all simulations, and the neglect of trip generation issues make the conclusions of this set of studies difficult to assess. In particular, their results tend to follow by assumption and so cannot inform policy.

A more complex series of simulations used a metropolitan planning authority's traffic impact model to consider how alternative future patterns of transportation investments and land-use patterns might affect the Portland, Oregon, region (1000 Friends, 1996). This is an important study because such exercises, involving integrated transportation and land-use models, are often used by regional planning and transportation agencies to evaluate alternative investment strategies. At the same time, most alternatives analyses rarely focus on the role of alternative land-use patterns.

The three primary alternative scenarios in this instance are a "no build" benchmark, which adds one new light rail transit (LRT) line but otherwise assumes no changes to land use or previously approved road

Table 3.2: Definition of Portland Alternatives

	Transportation Alternatives		
Mode	No Build	Highways Only	LUTRAQ
Land use	Existing plans	Existing plans	TOD
Transit	One new LRT line with feeder buses	"No Build" plus another LRT line and an express bus route	"No Build" plus four new LRT lines and four express bus routes
Roads	Only previously funded projects	A major bypass and 48 other improvements	Selected improvements; no bypass
Walk/bike	Existing	Existing	Existing plus improvements in TODs and LRT corridors
Demand management	None	None	Parking charges plus transit passes for workers

LUTRAQ = land-use/transportation/air quality; LRT = light rail transit; TOD = transit-oriented development.

Source: 1000 Friends, 1996.

plans; a "highway only" option, which adds a major highway and another LRT line; and a "land use/transportation/air quality" or LUTRAQ, option, representing a combination of higher residential densities, other transit-oriented development features, several additional LRT lines, higher parking costs, and subsidized transit passes for commuters. The alternatives are summarized in table 3.2.[4]

These scenarios were run through a metropolitan planning model, calibrated to the Portland area. That is, the simulations are essentially forecasts based on past behavior together with additional assumptions regarding trends in area demographics, the travel impacts of new roads, LRT lines, bus routes, parking charges, and transit subsidies. The key results are summarized in table 3.3. The main difference is that the LUTRAQ alternative doubles the mode share for commuting trips by transit. Trips and VMT for cars drop accordingly.

Above all, the Portland LUTRAQ simulations make this argument: Higher population densities near transit corridors for subsidized transit will increase the transit share of work trips. No doubt this is true. As travel by alternative modes becomes easier and less expensive, and travel by car becomes more costly, there will be migration from the latter to the former.

However, it is the *extent* of change that is the central question, and the LUTRAQ estimates of change are quite large. They are in turn based on estimates of ridership, trip generation, and VMT in Portland and other areas considered comparable, then adjusted further for the specifications of the alternatives in table 3.2. Thus, it is the accuracy of those estimates on which the simulations depend, in addition to the details of the alternatives themselves.

If residential densities increase in Portland along a transit corridor, how will transit ridership respond? If transit passes are subsidized, how will commuters respond? If parking becomes more expensive, how much less will drivers drive? These questions are not answered by the simulation; rather, they are inputs. The results in table 3.3 take these relationships as given, but they are not. The source for estimates in models of this sort are discussed under Multivariate Statistical Studies, below.

Descriptive Studies

Descriptive studies have the strong advantage of working from actual behavior. Their weakness is that, as with simulations, they do not attempt to explain that behavior. Worse yet, from the perspective of one interested in policy design, they attempt to explain very little at all. As such, descriptive work can only provide a simple accounting of travel experiences, individually or on average. This simplicity may well mask important interactions among the factors that explain such behavior. Two neighborhoods might exhibit different travel patterns, but this information is rarely sufficient to explain why those patterns are different.

Table 3.3: Simulated Transportation Impacts of Portland Alternatives

	Transportation Alternatives		
Travel Measure	No Build	Highways Only	LUTRAQ
Home-based work trip mode choice			
Walk/bike	2.8%	2.5%	3.5%
Single occupant vehicle	75.8%	75.1%	58.2%
Carpool	14.0%	13.6%	20.1%
Transit	7.5%	8.8%	18.2%
Total home-based mode choice			
Walk/Bike	5.1%	4.9%	5.6%
Auto	85.6%	85.4%	81.4%
Transit	9.3%	9.7%	12.9%
Total daily vehicle miles of travel (VMT)			
Daily VMT	6,883,995	6,995,986	6,442,348
% change from No Build		1.6%	−6.4%

LUTRAQ = land use/transportation/air quality.

Source: 1000 Friends, 1996.

Table 3.4: Travel Characteristics of Selected Communities Based on Travel Survey Data

Community	Vehicle Trips per Person per Year	Vehicle Trips per Household per Year (estimated)	VMT per Person per Year	Auto Driver Mode Share
Downtown San Francisco	210	481	1,560	NA
San Francisco	555	1,610	2,600	40%
Berkeley	695	1,800	3,300	45%
Oakland	660	1,709	4,160	55%
Daly City	730	1,898	5,500	59%
Walnut Creek	900	2,376	6,940	66%
Toronto	520	NA	NA	NA
Central City	NA		1,740	
Outer Suburb	NA		3,800	

VMT = vehicle miles traveled; NA = not available.

Source: JHK & Associates, 1996.

On the other hand, descriptive studies are an extremely important part of the process of understanding what is going on. They provide a picture, often very clear, of observed behavior and may contain important data and revealing insights regarding travel patterns in different settings. An example is table 3.4, compiled from various sources for a report prepared for the California Air Resources Board (JHK & Associates, 1995). Although table 3.4 does not tell us much about the differences in these cities, it is useful and interesting to see hard data on the range of trip generation rates, mode share, and VMT by location. In this set of cities, San Francisco and nearby yet suburban Walnut Creek are the outliers—and the gaps between them are impressive.

But these data must be interpreted with care. San Francisco and Walnut Creek have a multitude of differences, only partly due to land use and design features. The dangers of ignoring this fact are evident in another study frequently used to document the transportation merits of traditional or neotraditional street patterns. Working from household travel surveys from the San Francisco Bay Area, Friedman, Gordon and Peers (1992) categorized their observations into either "standard suburban" or "traditional," depending on whether each area possessed a hierarchy of roads and highly segregated land uses (the former) or had more of a street grid and mixed uses (the latter).

They then compared travel behavior in the two groups. Average auto trip rates were about 60 percent higher in the standard suburban zones for all trips, and about 30 percent higher for home-based nonwork trips. Just as in the cities in table 3.4, it is impossible to separate out the relative importance of the many differences between the two groups of communities in this format, however, and thus to identify how much of the observed behavior is influenced by the street configuration or any specific design feature alone. The traditional areas include those with employment and commercial centers, and with close proximity to transit networks servicing major employment centers, such as downtown San Francisco and Oakland. The standard suburban areas have lower densities, higher incomes, and longer commutes.

It is difficult to say what these results can tell us about the influence of any one feature, or any combination of features, without controlling for the many other significant differences among these communities. For example, in a descriptive examination of data from the 1990 National Personal Transportation Survey, Dunphy and Fisher (1996)

> confirmed the patterns found by other researchers of higher levels of transit use and lower automobile travel in higher density communities. However, the pattern is not as clear cut because of the intervening relationship between density and the demographic characteristics of certain households. For the national data and the individual regions examined, the current residents of higher density communities tend to be those with lower auto needs and greater transit dependency. (p. 90)

Rutherford, McCormack, and Wilkinson (1996) summarize actual travel behavior, using somewhat more detailed individual level travel diary data, and attempt to draw conclusions regarding how well behavior corresponds to various land use and design characteristics. Their interest is mainly with the influence of mixed land uses on weekend and weekday travel, and they employ a data set collected specifically for that purpose in the greater Seattle area. Travel diaries for three neighborhoods, two of them mixed-use, were compared with similar data for King County generally. Simple comparisons of average behavior in each neighborhood and the county reveal differences in mode choice, trip purpose, trip chaining, trip chain lengths, transit mileage, and VMT. The authors conclude that their information

> generally supports the notion that mixed-use or neotraditional neighborhoods can reduce the amount of travel for most households . . . although we concur with others that the linkage is very complex. Residents of the two mixed-use neighborhoods in Seattle traveled 27 percent fewer miles than the remainder of North Seattle, 72 percent fewer than the inner suburbs and 119 percent fewer than the outer suburbs. (p. 54)

The study does acknowledge that these neighborhoods differ in several respects, such as age, labor force participation, and income, but the nature of the analysis does not permit a formal examination of the roles of those differences.

Again, the evidence is consistent with the idea that people in mixed-use neighborhoods travel differently, but it neither demonstrates that the mixed-use character of the neighborhood is responsible nor establishes that reducing the land-use homogeneity of suburban neighborhoods would change residents' travel behavior.

The studies reviewed in the next section try a different approach, one that in principle can address these and other methodological challenges more directly.

Multivariate Statistical Studies

Studies in this category also examine observed, rather than hypothetical, behavior. In addition, their attempts to *explain* rather than merely describe what is going on are on more solid methodological footing. Still, this remains a challenging task given the many reasons people have for choosing to travel as they do; it is also a key step in understanding the manner in which planning and design strategies influence driving and other travel outcomes. This is the primary reference literature for the questions of this book.

The studies in this category vary in several significant ways. First, they ask different questions of their data. Second, their data capture different features of the built environment and of travelers, and at different levels of detail. Third, they investigate their data by various means.

The complexity of travel behavior, together with the difficulty of isolating and explaining the role of individual features of the built environment, indicates the need for an analytical method that controls for as many differences among circumstances and behaviors as are necessary. This would permit the analyst, ideally, to test the specific hypothesis that a particular urban design element influences travel in one direction or another and at a certain magnitude, while controlling for the independent influence of household income, travel demands, mode availability, and the like.

Multivariate regression analysis fits the bill quite well, although the appropriateness of the method and the credibility of the statistical results in turn depend on a good number of other critical assumptions regarding the form of the data and the structure of the underlying behavior (Greene, 1993). It is not enough, in other words, to have good measures of all the factors in question and then to regress an observed travel outcome on them. The two most critical sets of assumptions concern the *specification* of the regression (which variables are to be included and in what manner), and the *estimation* of the regression (which statistical procedure is appropriate to the form of the data and relationships among the variables). In addition, there may be limits to what one can learn from aggregate data, for example, particularly where resources, constraints, demographics, land-use patterns, and other factors vary considerably among travelers and places.

As indicated earlier, we divide this literature roughly into two parts.[5] In the first, the relationship between travel outcomes and urban form variables is significantly ad hoc in that it lacks a strong or even clear behavioral foundation. These studies may be based on a description of a choice process, say, where the factors influencing the relative attractiveness of alternative travel modes are discussed, perhaps at length and in detail. This label is not offered pejoratively, but only for lack of a better term to refer to analyses with no explicitly systematic theory of choice, or model, of how decisions among options are made in a system of exogenous and endogenous environmental factors (see, e.g., Kreps, 1990).

In the second group of studies, the selection of variables and estimation procedure are motivated, usually, by an explicit behavioral framework. Still, the dividing line is not a hard one, and some studies belong in both, or perhaps in neither. We hope the distinction and subsequent discussion is useful as an organizing scheme nonetheless.

Ad Hoc Models

Improved data and statistical procedures in recent years mean that the studies in this category are generally both thoughtfully constructed and informative. They consider many measures of urban form while attempting to control for differences among communities, neighborhoods, and travelers. At the same time, however, the travel decision process is neither well developed nor explicitly described.

Handy (1996a) examined travel diary data for two pairs of cities in the San Francisco Bay area. She found some differences in nonwork trip frequencies associated with differences in local and regional shopping opportunities. In this instance, neighborhoods are categorized and indexed by accessibility measures such as blocks per square mile, cul-de-sacs per road mile, commercial establishments per 10,000 population, and accessibility to retail centers. The differences, when statistically significant, suggest that neighborhoods that are closer to shopping destinations generated more trips, raising the possibility that increased accessibility—measured as a combination of proximity, density, and street pattern—might increase rather than decrease trips. Her results also suggested that the effects of neighborhood design are greater than the effects of household characteristics when comparing time, frequency, and variety of trip destinations among the traditional and suburban neighborhoods.

Cervero and Gorham (1995) examined matched pairs of communities selected to juxtapose "transit-oriented" land-use patterns with more typical post–World War II developments. They compared work and nonwork trip generation rates for seven pairs of neighborhoods in the San Francisco Bay area and six pairs of neighborhoods in the greater Los Angeles metropolitan area. Neighborhoods ranged in area from one quarter square mile to two and a quarter square miles. This relatively small geographic scale (not much larger than census tracts) is typical of virtually all recent empirical work on this topic, and the small geographic scale is also true to the neighborhood scale emphasized in many recent proposals.

They hypothesized that transit-oriented neighborhoods generate more pedestrian and transit trips. These neighborhoods were identified using street maps, transit service information, and census data describing median household income. The travel data came from census data describing the journey to work, summarized by census tract. The authors suggest that street layouts do influence commuting behavior—transit neighborhoods averaged higher walking and bicycling modal shares and generation rates than did their automobile counterparts. However, this finding held only for the Bay Area neighborhoods. In the Los Angeles–Orange County comparisons, the differences in the proportion of transit or pedestrian trips between the transit- and automobile-oriented neighborhoods were negligible. Cervero and Gorham suggest the sprawling nature of the region explains the weaker results for the Los Angeles–Orange County comparisons. In some ways, the potentially dominant role of the surrounding regional circulation pattern is a difficult hurdle for proponents of neighborhood-scale solutions to traffic problems. (Handy [1992] and McNally [1993] address this issue explicitly.)

Holtzclaw (1994) measured the influence of neighborhood characteristics on auto use and transportation costs generally. The neighborhood characteristics used in the study are residential density, household income, household size, and three constructed indices: transit accessibility, pedestrian accessibility, and neighborhood shopping. These are in turn used to explain the pattern of two measures of auto use: the number of cars per household, and total VMT per household. The data are from smog-check odometer readings and the 1990 U.S. Census of Population and Housing for 28 California communities. The reported regression coefficient on density in each case is −0.25, suggesting that doubling the density will *reduce* both the number of cars per household and the VMT per household by about 25 percent.

The results also argue that a doubling of transit accessibility, defined as the number of bus and rail seats per hour weighted by the share of the population within a quarter mile of the transit stop, will reduce the number of autos per household and the VMT per household by nearly 8 percent. Changes in the degree of pedestrian access[6]—based on street patterns, topography, and traffic—or neighborhood shopping had no significant effect on the dependent variables in this sample, however.

Yet the results from Holtzclaw (1994) are based on weak statistical analysis. The regressions include, as independent variables, only a small number of the variables mentioned above. For example, the result for automobile ownership is based on a regression of household car ownership rates on one variable—residential density. This approach highlights correlations between pairs of variables, but hypothesis testing and causal inference is obscured. The end result is an assessment of how VMT and automobile ownership vary with density without explaining much of the causal structure that links those with other variables.

Kulkarni (1996) examined 1991 travel diary data for twenty neighborhoods in Orange County, California. The neighborhoods were classified as traditional neighborhood developments (reflecting land-use patterns consistent with Neotraditional or New Urbanist designs), planned unit developments (characterized by separated land uses and curvilinear street patterns), and an intermediate or mixed case. The traditional neighborhoods generated the fewest trips per household, and the planned unit developments generated the most trips per household, but once income differences across neighborhoods were controlled (in an ANOVA analysis), income proved to be a much better predictor of differences in trip generation across neighborhoods.

Messenger and Ewing (1996) provide an interesting attempt to isolate the independent effect of land-use mix and of the street network by accounting for the joint decision to travel by bus and to own a car. They use 1990 Census data at the traffic analysis zone level for work trips in Dade County, Florida, and thus do not model individual decisions. Still, they find that density affects the share of zone work trips by bus only through its affect on car ownership. Again, the relationship between density and travel behavior appears too complex to be reduced to a simple design criterion.

Two important methodological shortcomings are apparent in most of these studies. First, in examining associations between neighborhood type and aggregate measures of travel behavior, it is crucial to disentangle the effect of urban design and land use from the effect of systematic demographic differences across neighborhoods. Do residents in dense neighborhoods travel less because of the density of their neighborhood, for example, or do dense neighborhoods attract people who prefer not to travel by car? The policy implications of this distinction can be crucial, as illustrated by Kulkarni (1996). He suggests that the statistically significant association between neighborhood type and car trip rates is, more properly, an association between household incomes and car trip rates. This raises the possibility that neighborhood designs might have little impact on travel behavior unless incomes somehow vary from design to design. In new neighborhoods, with above-average housing costs, this is of course quite feasible.

Second, the relationship among neighborhood attributes, the characteristics of residents, and travel behavior is complex. Many of the relationships that must be understood for policy analysis are obscured by aggregate data. Similarly, behavioral models of travel are best specified and fitted on individual or household level data, since those are the decision-making units. Regression analyses of individual travel data hold the promise of overcoming the shortcomings of statistical studies of aggregate data. We describe studies using individual data below. While often innovative, they still lack a clear behavior framework.

Cervero and Kockelman (1997) and Kockelman (1997) use travel diary data for persons in 50 and 1,300 San Francisco Bay Area neighborhoods, respectively, to examine the link between VMT (per household), mode choice, and land use near a person's residence. The neighborhoods were chosen to correspond to either one or two census tracts. VMT and mode choice were regressed on a set of individual sociodemographic variables and variables that included population and employment densities, indices of how residential, commercial, and other land uses are mixed in close proximity, and street design data for the person's residential neighborhood. The land-use variables had a significant effect in some of the models, but the elasticities implied by the regression coefficients were often small compared with sociodemographic variables.

A 1993 study of Portland, Oregon, is similar in approach to the Holtzclaw report, but has the advantage of using household-level survey data (1000 Friends of Oregon, 1993). This analysis also attempts to explain the pattern of VMT, as well as the number of vehicle trips, using household size, household income, the number of cars in the household, the number of workers in the household, and constructed measures of the pedestrian environment, auto access, and transit access. The auto and transit access variables were defined as simple measures of the number of jobs available within a given commute time: 20 minutes by car and 30 minutes by transit. As an example, an increase in 20,000 jobs within a 20-minute commute by car is estimated to reduce daily household VMT by half a mile while increasing the number of daily auto trips by one-tenth of a trip. The same increase in jobs within a 30-minute commute by transit reduced daily VMT a bit more, to six-tenths of a mile, and reduces the number of daily car trips by one-tenth of a trip.

The pedestrian access variable is more complex, based on an equal weighting of subjective evaluations of four characteristics in each of 400 zones in Portland: ease of street crossings, sidewalk continuity, whether local streets were primarily grids or cul de-sacs, and topography. The final score for each zone ranges from a low of 4 to a high of 12, with 12 being the most pedestrian friendly. The regression model reported that an increase of one step in this index, from 4 to 5, say, decreases the daily household VMT by 0.7 miles, and decreases the daily car trips by 0.4 trips. These point estimates are used to predict the effects of changes in the independent variables, such as access to employment by transit, on the dependent variables. Although this result is consistent with the idea that neighborhood features influence travel, the composite construction of the pedestrian access measure limits its usefulness for policy. Since the effects of the street pattern are not separated out from the sidewalk, street crossing, and topography variables, we cannot say which features matter the most, or if each matters individually or only in tandem with others.

Kitamura, Mokhtarian, and Laidet (1997) add data on personal attitudes to the list of explanatory variables. Travel diary data for persons in five San Francisco Bay Area neighborhoods were regressed on sociodemographic variables, land-use variables for the person's residence, and attitude variables that were drawn from survey responses designed to elicit opinions on driving, the environment, and related questions. (The five neighborhoods averaged approximately a square mile in area.) The idea is to consider the relative contribution that attitudes have on travel behavior beyond land use or neighborhood characteristics.

The authors first regressed socioeconomic and neighborhood characteristics against the frequency and proportion of trips by mode. High residential density was positively related to the proportion of nonmotorized trips. Similarly, the distance to the nearest rail station and having a back yard were negatively associated with the number and fraction of transit trips. But, as the authors ask, do people make fewer trips because they live in higher density neighborhoods, or do they live in higher density areas because they prefer to make fewer trips? The attitudinal measures (including attitudes toward various residential and travel lifestyles) entered significantly, and appeared to explain behavior better the land-use variables (see also Kitamura et al., 1994). However, the analysis is only a first stab at accounting for preferences in travel behavior models. It does not, for example, model the relationship between preferences and locational choice.

Cervero (1996) is mainly interested in how work trip mode choice is affected by the land-use mix. He used individual level data on eleven metropolitan areas from the 1985 American Housing Survey, which includes data on the density of residential units and the location of nonresidential buildings in the vicinity of the surveyed household. The model estimates the probability of choosing a given travel mode for the commute as a function of land use variables (type of housing structure within 300 feet, commercial or other non-residential building within 300 feet, grocery or drug store between 300 feet and one mile), a dummy indicating if the household lived in the central city, the number of cars available to the household, the adequacy of public transportation, and the commute length.

His results suggest that people were less likely to drive to work, and more likely to use transit, if commercial or other nonresidential units were nearby, if nearby housing was medium to high density, if they lived in the central city, if they had short commutes, and if they had few cars. This is consistent with the idea that commuters are more likely to use transit if they can stop to shop, and so on, on the way home from the transit stop. The effects of higher densities and car ownership were stronger still. A two-stage car ownership model, where the commute length is treated endogenously, and a two-stage commute length model, where car ownership is endogenous, give similar results. In both cases neighborhood residential density and central city location have significant negative effects on the probability of owning a car and commute length.

Handy, Clifton, and Fisher (1998) examine pedestrian trips for two purposes, strolling and shopping, based on survey data they collected from selected Austin, Texas, neighborhoods. The report emphasizes the importance of qualitative analysis of their survey data, indicating the complexity of accounting for pedestrian travel behavior and attitudes, but it also includes an interesting statistical model that regressed the number of walking trips on socioeconomic variables (age, employment status, children under age 5 in the home, gender, and categorical measures of income) and within-neighborhood urban form variables (perception of safety while walking, shade coverage, how interesting the local housing is, scenery provided by trees and houses, level of traffic, and frequency and desirability of seeing people while walking). In addition, the strolling model included a dummy variable for whether or not the person walked a pet, and the store model included variables measuring the distance to a store, ease of walking, and walking comfort.

Among the urban form variables, only perceived safety, shade, and the

Figure 3.4. Pedestrians on Broadway and 5th, downtown Los Angeles, in the early '30s. (Security Pacific Collection, Los Angeles Public Library.)

"people" variable significantly explained strolling trips, while the housing and scenery variables were significant in the store trip variables. Three cost variables in the store model are distance, ease of walking, and walking comfort; all were significant with the expected signs.[7]

A comparison of these studies reveals many differences in travel outcome variables, independent variables, statistical approach, and results. For example, Holtzclaw (1994) and 1000 Friends (1993) offer evidence that higher density, more accessible neighborhoods are associated with fewer cars and VMT per household, and with lower car trip rates. Yet Handy (1996a) reports that neighborhoods that are closer to shopping destinations are associated with more shopping trips, and the results of both Kulkarni (1996) and Kitamura, Mokhtarian, and Laidet (1997) suggest that relationships between travel outcomes and neighborhood characteristics might be driven by often unmeasured independent demographic characteristics and attitudes.[8] These unmeasured factors can affect the policy implications of this literature.

Given such variation in results and messages, one might be tempted to count studies that support a given conclusion and argue from a preponderance of the evidence—as Ewing (1997a) and Burchell et al. (1998) have. Yet that would be shaky given the evidence that a study's results might vary with the pattern of regional accessibility (Handy, 1992; Cervero and Gorham, 1995), individual characteristics and attitudes (Kulkarni, 1996; Kitamura, Mokhtarian, and Laidet, 1997), or assumptions regarding how variables should be measured, what should be included in the statistical model, and how the statistical models should be and can be estimated. In short, a summary of this literature must include a comparative assessment of the methodological quality of the various studies and thus of the reliability of their results.

Yet it is hard to summarize the ad hoc statistical literature reviewed above succinctly for at least two reasons: the absence of a systematic choice theory, to help identify how specific hypotheses regarding urban form relate to the rationality of travel behavior, and the subsequent difficulty of comparing one study's results with another's. The point of departure for the next section is the argument that the literature on the transportation impacts of urban form have rarely employed a strong conceptual framework when investigating these issues, making both supportive and contrary empirical results difficult to compare or interpret. In particular, an analysis of trip frequency and mode choice requires a discussion of the *demand* for trips, but this is often lacking in planning and land-use studies at even a superficial level. That approach should permit us to explore the behavioral question, for example, of how a change in trip distance influences the individual desire and ability to take trips by various modes.

A demand framework outlines how overall resource constraints enforce trade-offs among available alternatives, such as travel modes or the number of trips for different purposes, that is, how the relative attractiveness of those alternatives in turn depends on resources and relative costs, such as trip times and other expenses. The studies summarized next are either explicit or implicit in their use of this approach.

Demand Models

As mentioned above, the travel demand literature is extensive and methodologically advanced (for surveys, see Train, 1986; Small, 1992; Small and Winston, 1999). However, urban form and land-use factors are typically ignored. The travel demand literature that does consider urban structure and design is mainly concerned with the journey to work. The studies reviewed in this section include both land-use and conventional demand variables, such as unit travel costs, income, and taste controls, whether or not the authors specify a full-blown demand model. In other respects, however, the analyses are less sophisticated in key respects than are studies we have characterized above as ad hoc. Again, this categorization is a labeling convention only.

To begin, we look at Kain and Fauth (1976), one the earliest studies to use disaggregate data to explain urban travel behavior as a function of both economic circumstances and urban form. As the authors put it, "this study seeks to determine how the overall arrangement of land uses, the density, location, juxtaposition of workplaces and residences, in combination with the transit and highway systems serving them, affect the level of auto ownership and mode choices of urban households" (p. 15).

Using 1970 Census individual level travel data from the largest 125 Standard Metropolitan Statistical Areas (SMSAs), they estimate work trip mode choice models that in turn use the results from regression models of auto ownership estimated in earlier stages. Their urban form data include measures of central city density, central business district (CBD) employment, the percentage of the housing stock that is single family, workplace location (CBD, central city, or suburb), and the supply of highway and transit services in each SMSA. In addition, these models are explicitly configured as demand models, although several important demand variables, such as the cost of auto ownership and the relative costs of travel by each mode, are either left out or assumed to be captured by urban structure measures.

Although the sample was limited to White, one-worker households, several results are interesting. Most of the variation in the mode choice models is explained by the car ownership equations. This result appears in other work as well, and underscores the importance of the car in travel behavior, apart from other elements of the travel environment (cf. Messenger and Ewing, 1996; O'Regan and Quigley, 1999). The value of the Kain and Fauth (1976) study lies in part, then, in the explanations it offers for why these households have cars. They find that "differences in the level of transit service, parking charges, and workplace and residence densities play a larger role in determining the level of auto ownership in

CBD than in non-CBD workplaces" (p. 47). The presence of a rail transit system affected car ownership in all cases, while the bus service variable did not. The residential density variable also significantly influenced car ownership, with a particularly pronounced effect on the probability of not having a car for both CBD and non-CBD workers. On the other hand, CBD or central city workers in households with two or more cars drove more than their lower density counterparts.

As an illustration, they applied the models to a comparison of the behaviors of Boston and Phoenix residents, who had roughly the same average socioeconomic characteristics (Kain and Fauth, 1977). There was no difference in the proportion of households owning one car in the two places. However, they calculated that differences in urban form—as measured by the age of the housing stock in each county, the percentage of the area's units that are single family, and the density of the structure in which the household lives—explained nearly two-thirds of the difference in the proportion of households without cars in these two regions in 1970. Thus, the study does provide evidence that urban form matters, though mainly as a determinant of car ownership. In turn, once people have access to cars, they tend to drive to work regardless of where they live or the structure of their community.[9]

Kain and Fauth (1976, 1977) removed non-White households from their sample in order to avoid analyzing differences by race, which they anticipated would involve additional market problems due to discrimination. However, the "spatial mismatch" literature was founded by Kain (1968) and is primarily concerned with racial differences in *choices* regarding the journey to work. Blacks are typically, though not always, found to face longer commutes or fewer employment opportunities near their homes than Whites. This is frequently taken as evidence that the choices of the former are constrained relative to the latter (Ellwood, 1986; Gordon, Kumar, and Richardson, 1989c; Kasarda, 1995; O'Regan and Quigley, 1998). (Taylor and Ong [1995], using American Housing Survey data, found commutes by Blacks to not be longer in distance.)

One explanation is housing discrimination, limiting the ability of Blacks to live closer to suburban jobs, and another is differences in car ownership rates.[10] Or, as O'Regan and Quigley (1998) put it

> In sum, two primary forces are responsible for the specific link between transport access and employment which limits the economic opportunities available to low-income and minority households—slow adjustment in real capital markets to changes in locational advantage and explicit barriers to the residential mobility of low-income or minority households. (p. 9)
>
> So, while only 11.5 percent of households nationally are without an auto, 45 percent of central city poor black workers and 60 percent of central city poor black nonworkers have no access to a car. (p. 30)

Although this work reveals some interesting interactions between mode use and commute length typically ignored by the design literatures, with transit users experiencing considerably longer commute times, none of these studies include variables capturing the effect of urban structure beyond the decentralization of employment and population.

The role of urban structure is explicitly considered by Giuliano and Small (1993). They use 1980 journey-to-work data for the Los Angeles Consolidated Metropolitan Statistical Area, a region of 10.6 million persons and 4.6 million jobs at the time, for 1146 geographic units known as travel analysis zones. These data include estimates of inter- and intrazonal distances and peak travel times. From these they calculated the minimal required commutes by zone to each of the many employment centers and subcenters based on the local jobs/housing balance. Notably, required suburban commutes are shorter than those of people working downtown and only one-third to one-quarter as far as actual commutes. Thus, density is *inversely* related to the required commute length. Both travel costs and jobs/housing balance appear to matter when explaining commuting distances and times, but not much. They conclude that policies attempting to change the metropolitanwide land-use structure will have disappointing impacts on commuting.

Shen (1998) recently revisited this approach for 787 traffic analysis zones in the Boston metropolitan area. Though not an explicit demand analysis, his study includes many elements of one. Rather than utilize measures of jobs/housing balance and the minimal required commute (as calculated by an assignment model) to represent urban structure, he adopted the "accessibility" literature strategy of using a gravity formulation to measure access to employment (see note 5). This can be interpreted as an average travel price of sorts.

Shen's (1998) measure is a weighted score of the travel times between workers' homes and jobs that accounts for car ownership rates. The demand variables are limited to income, poverty status, and the accessibility measure, as a weighted index of travel cost, which doubles as the urban structure variable. He then regressed 1990 commute times, from the Census, on these and household traits, mode, and occupational variables. Shen interprets the result that greater access is significantly associated with less commuting as evidence that the land-use/transportation linkage still matters, weak though it may be.

A recent dissertation by Kockelman (1998) makes progress on several fronts. First, her modeling of travel choice is explicitly derived from modern demand theory. In addition, her treatment of urban form and land use is extensive, incorporating the following measures for the San Francisco Bay area in 1990: accessibility to all jobs by automobile, accessibility to sales and service jobs by walking, mix of neighborhood land uses, mix of neighboring land uses, and developed-area densities (as in Kockelman [1997], which does not employ a demand model). A key modeling strategy is to treat travel times and costs as choice variables rather than parameters. She then uses these urban form measures as instruments for the nonwork travel times and costs associated with different locations, after controlling for trip purpose/activity type. These first-stage regressions do not perform well, however, and the individual coefficients on the variables are not reported. That is, Kockelman (1998) estimates trip lengths as a function of urban form, but only to obtain an estimated trip length as the first stage of later models of the number of trips for different purposes, her focus. Urban form does not enter the trip demand models directly.[11]

Demand studies of the influence of urban form on travel have some appeal, given their attention to such basic issues as travel costs and behavioral trade-offs. Much progress remains to be made, however. The missing step seems to be the consistent and explicit linkage of individual urban form and land-use measures to the economic concepts of price, cost, and quality.

Summary

The results described in this chapter are mixed and messy. Numerous studies report that higher densities, mixed-use development, more open circulation patterns, and "pedestrian-friendly" environments are all associated with less car travel. The data are often poorly suited to the purpose, the research designs are faulty or ad hoc and thus difficult to generalize, and the statistical methods applied to the data are typically primitive. This does not mean that the results are incorrect, only that they may lack sufficient robustness to be the basis for policy.

The research strategy in most empirical analyses is simply to search for correlations among neighborhood features and observed travel—sometimes controlling for other relevant factors, sometimes not. Interpreting the range of results in any one case is also problematic because the causal theory is not clearly established. What can we generalize about the factors that generate more car trips in one environment and less in another? Is density a proxy for demographics, distance, car ownership rates, transit service levels, and so on? While some studies based on observed behavior do attempt to control for different trip purposes (e.g., shopping vs. commuting), trip lengths (e.g., neighborhood vs. regional), and demographic variables likely associated with trip demand (e.g., income, gender, and age), the approach is commonly idiosyncratic. Further, the wide range of outcomes found in this work reveals little about whether a particular land-use pattern or urban design feature can deliver the reported transportation benefits.

The challenge is that empirical work of this nature is problematic given the enormous complexity of the behavior to be explained and the great difficulties of conceptualizing the interaction of travel and the physical form of the city.

The next two chapters address the drawbacks identified in this chapter in two principal respects. Chapter 4 sketches out a very simple but consistent choice framework. Our purpose is to clarify in a transparent manner how travelers and potential travelers likely weigh travel costs against the value of travel when making decisions about how many trips to take. We next discuss what these elementary behavioral foundations imply about how alternative land-use and urban design features might influence travel choices in general as a way to identify which behaviors follow clearly from design, which do not, and which may or may not. That is, we set up a choice framework to help us identify straightforward hypotheses. The concluding part of chapter 4 adapts that framework for empirical testing, which takes place in chapter 5.

To address several issues raised in chapter 3, in this chapter we develop a conventional behavioral story. The relative attractiveness of options regarding where to go, how often to go, and how to get there likely depends, at least in part, on how their costs and benefits compare. After explicitly including urban form and land-use variables as characters in the story, we describe their general relationships. In the end, we argue that certain travel outcomes follow from alternative urban forms directly, while other outcomes cannot be identified without data. Then we propose a strategy for using data to test various travel hypotheses.

4 The Demand for Travel

As described in chapter 1, the new urban designs are part philosophy, part art, part economics, and part social optimism. Still, a key to their popularity is the open embrace of conventional and even conservative standards of neighborhood form, scale, and style. Many new urban designs self-consciously recall small town settings where neighbors walk to get a haircut and stop on the way to chat with neighbors sitting on the front porch watching the kids play. The attraction of these ideas is subjective, personal, yet pervasive. After all, in principle, what is not to like about pretty homes in quiet, friendly, and functional neighborhoods?

But will they improve the traffic? Chapter 3 concluded that existing evidence is unsatisfactory in several respects. Among the problems identified in the literature was the common absence of a conceptual framework for hypothesizing how urban form might be expected to influence travel behavior. In particular, only a small share of the studies in this area even attempt to model travel behavior in the conventional manner, that is, as travel demand.

In this chapter, we develop a framework for consistently evaluating the net travel impacts of changing land-use patterns, such as many new urban designs propose. The idea is to adapt a simple model of travel demand to measurable urban form elements. This permits us to derive specific conclusions that follow directly from the assumptions of the model as well as specific hypotheses that can be tested only with data on observed behavior. These assumptions are summarized in figure 4.2. The last part of the chapter develops an empirical implementation of the model and these hypotheses, which is applied to data in chapter 5.

A Behavioral Framework

The theory of demand provides perhaps the most straightforward way to analyze travel behavior, by emphasizing how overall resource constraints force trade-offs among available alternatives, such as travel modes and trip distances, and how the relative attractiveness of those alternatives in turn depends on relative costs, such as trip times. This

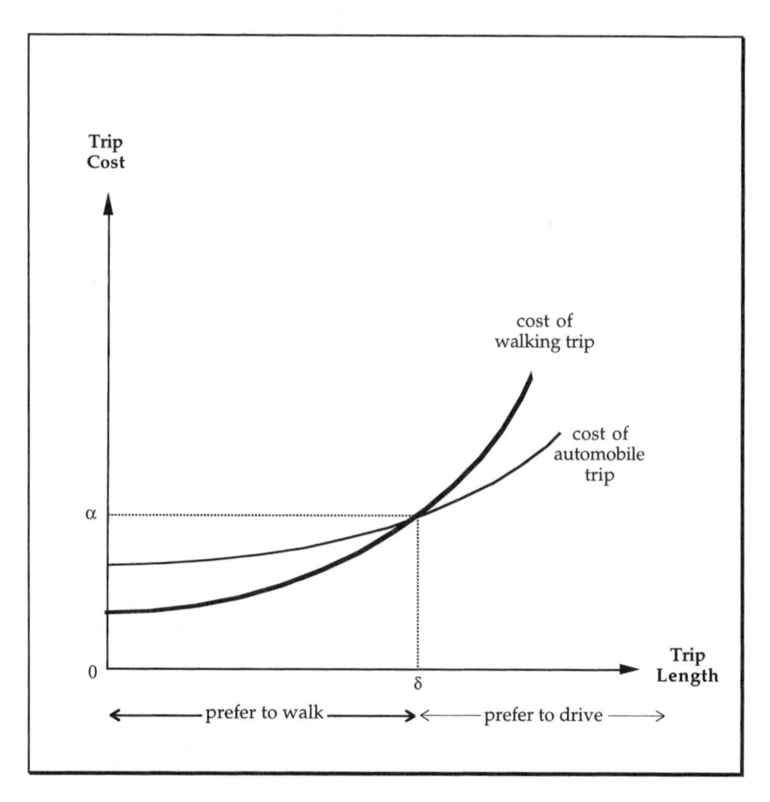

Figure 4.1. Walk or drive? A stylized comparison of the choice to walk or drive based on the costs associated with trip length—including time, effort, and out-of-pocket costs. In this particular example, trips on foot are initially less costly; walking is the least-cost mode for trips shorter than δ miles. Beyond δ miles, the automobile provides a less expensive trip overall.

framework assumes that individuals make choices, either alone or as part of a family or other group, based on their *preferences* over the goods in question, the *relative costs* of those goods, and *available resources* (e.g., Kreps, 1990). Preferences include attitudes and tastes, for example, regarding the experience of driving versus walking, and are likely correlated with demographic and other personal idiosyncratic characteristics. But the decision to take a trip to the coffee shop by car or by foot depends not only on how one feels about those options, but also on external factors over which one has no or only limited control, for example, on the cost of one mode versus the other—I may prefer to drive but if the gasoline or parking expenses of doing so are high enough, walking may appear to be the better choice. Thus, the demand for walking trips is explained not only by one's preferences across modes but also on the cost of walking relative to the cost of other modes.

The role of accounting for available resources is mainly to fix the importance of costs; the impact of a $5 parking charge on your decision to drive to the coffee shop depends on what funds you will have left for that double expresso you need to get you through the afternoon. Note the framework applies just as well to any situation where decisions are made concerning the allocation of scarce resources, whether or not they involve actual money. In the model presented below, for example, the scarce resource is time, and each mode is compared in terms of the time consumed rather than the cash. Note also that this framework does not explain preferences; it only explains how one makes informed decisions given those tastes together with costs.

The discussion below abstracts from the many other aspects of this topic to address the effect of improved access on travel distance, trip frequency, and mode split. Two sets of assumptions focus the analysis on the questions at hand:

1. "Access" is interpreted solely as a price or cost characteristic, related to trip length.[1]
2. New urban designs are assumed to reduce the distance required to make any local trip.

In a sense, the last assumption characterizes these designs as a compression of existing land-use patterns that, most particularly, shrink effective travel distances between potential nodes. Compared to an alternative design, this improvement in access has three somewhat countervailing effects: It reduces the *absolute* cost of a trip in each mode; it may change the *relative* cost of each mode; and it increases the purchasing power of any individual making that trip by freeing up time and money resources. Although the literature on the new urban designs tends to suggest otherwise, the first and third of these effects will typically increase the demand for trips in all modes rather than reduce it.[2] The second may or may not. The presumption would be that pedestrian travel could become more attractive in comparison with driving than it had been, possibly because travel distances for some trips can be made short enough to facilitate walking, or because pathways, streetscapes, and public spaces encourage more pedestrian activity.

Mode Choice

As benchmarks, the potential effects of the price changes on mode choice are illustrated in figures 4.2, 4.3, and 4.4 for trips by car and by foot.[3] For any given trip frequency, these figures plot the cost of a trip, for some unspecified purpose, against trip length. This cost summarizes all the relevant features of the trip, including the aesthetic aspects so critical to the Neotraditional planners. The purpose of the trip has obvious implications for the relative merits of walking and driving, and for how those merits vary with the length of the trip. (As often noted, people rarely walk to the grocery store when they can drive due to the weight of their return trip load.) Each chart assumes that the marginal cost of travel is everywhere rising; both the total trip cost and the marginal cost of walking are initially lower than for driving, and the cost of walking rises more quickly than that for driving does. Hence, people will tend to walk for short trips and drive for longer trips, all things considered. These idealizations are intended only to clarify how access can influence the means of travel.

Figure 4.2 presents an initial situation, wildly simplified for the sake of legibility. For short trips, walking is the preferred mode. When the cost (including the time required as well as out-of-pocket expenses) for the trip gets to a certain point, however, this person prefers to drive. In the example, that cost is labeled α and corresponds to a trip of length δ. For trips of distance δ or more, say, one-quarter of a mile, it costs less overall to drive and the car becomes the best mode. The lower envelope of the two total cost curves is the mode demand curve at any distance.[4] *Hence, any change in land-use patterns that reduces trip length enough, in this case from above* δ *to below* δ, *will substitute pedestrian for automobile traffic.*

If land-use patterns lower travel costs across the board, the relative attractiveness of driving versus walking will depend on the relative change in the cost of each. Figures 4.3 and 4.4 illustrate two such cases. The cost of traveling any given distance decreases for both modes in each example. An asterisk denotes the postimprovement trip cost, such that walking trips to any distance have fallen from a cost of w to w^*. In figure 4.3, the pedestrian cost falls the most at any distance, so the trip length where modes change (δ^*) becomes longer; that is, $\delta < \delta^*$. *For any given number of trips, the mode split now features more trips by walking and fewer by car than before.* This is consistent with the work on pedestrian travel by Untermann (1984), Guy and Wrigley (1987), 1000 Friends of Oregon (1993), and Handy, Clifton, and Fisher (1998), all of which show that walking trips rise with an improvement in pedestrian access.

This is not the only possible outcome, however. New urban designs also promise to improve circulation and reduce trip lengths for automobile travel, and designers have rarely if ever explicitly compared how these improvements compare with the value of the community's pedestrian-oriented features. It is possible that the grid circulation pattern characteristic of many New Urbanist designs could generate the result shown in figure 4.4, where a reduction in street congestion along with other changes lowers per-mile auto travel costs the most. In some instances, the change in automobile circulation is the focus of the design.

While many of the travel-oriented components of the new neighborhood designs are aimed at encouraging pedestrian and transit travel, they often also include changes in street patterns that will reduce the distances required to drive between locations. Will this lead to more walking and less driving, as promised? The charts above suggest that the net impact on mode choice is ambiguous, except where the (time and money) cost of nonauto modes are reduced sufficiently more than car travel.

Trip Generation and Vehicle Miles Traveled

What cannot be easily answered with these figures is the impact of improved access on total trip generation and thus on the total amount of

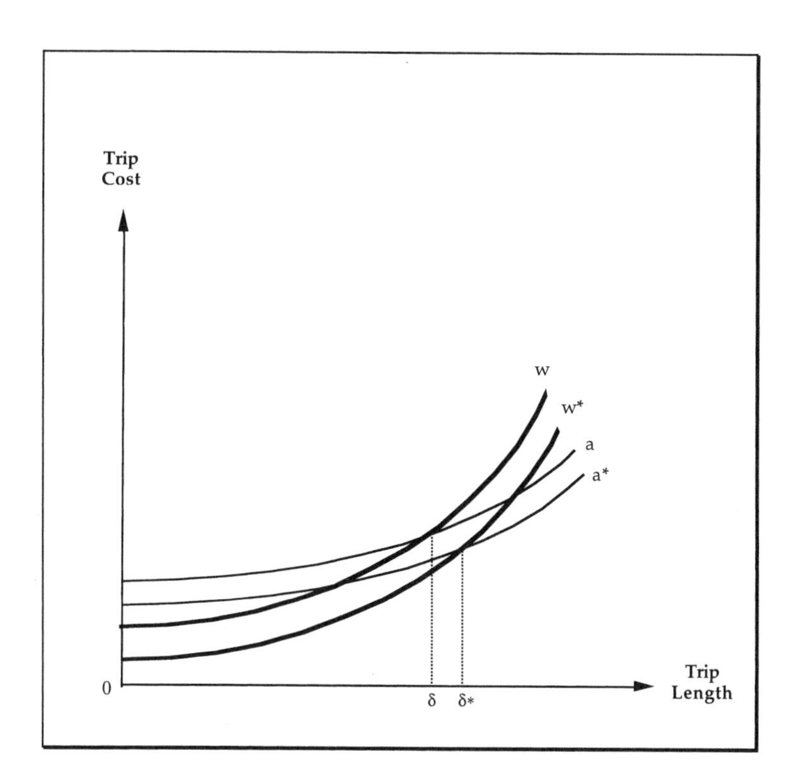

Figure 4.2. The new urbanism. The grid street pattern lowers trip costs for both modes in comparison to conventional designs. Walking costs fall from *w* to *w**, and automobile trip costs fall from *a* to *a**. In this example auto travel costs fall *less* than walking costs, so maximum walking trip length *rises* from δ to δ*.

travel by mode. Depending on how relative access changes, more trips are likely to be generated in certain modes, including possibly car travel. Even in those cases where better access translates to a shift from cars to pedestrian travel for preexisting trips, new trips by car may result in response to the lower cost per trip. *Whether total car travel—trip frequency times trip length—rises or falls therefore depends on how these two components compare.* If the number of automobile trips increases by more than the average trip length declines, a result opposite to the New Urbanist promise is obtained.

To focus on the behavior of interest, consider the problem of individuals making choices over five uses of time: the number of trips they complete by car, foot, transit, or some other transportation mode, and a composite good representing all other uses of time. (I.e., the model abstracts from other decisions, which is different from assuming they do not happen but does imply they are not a central feature of the story.) For most purposes, a trip is a derived demand, meaning that people typically travel as a means to an end, not as an end in itself. A "trip" is thus de-

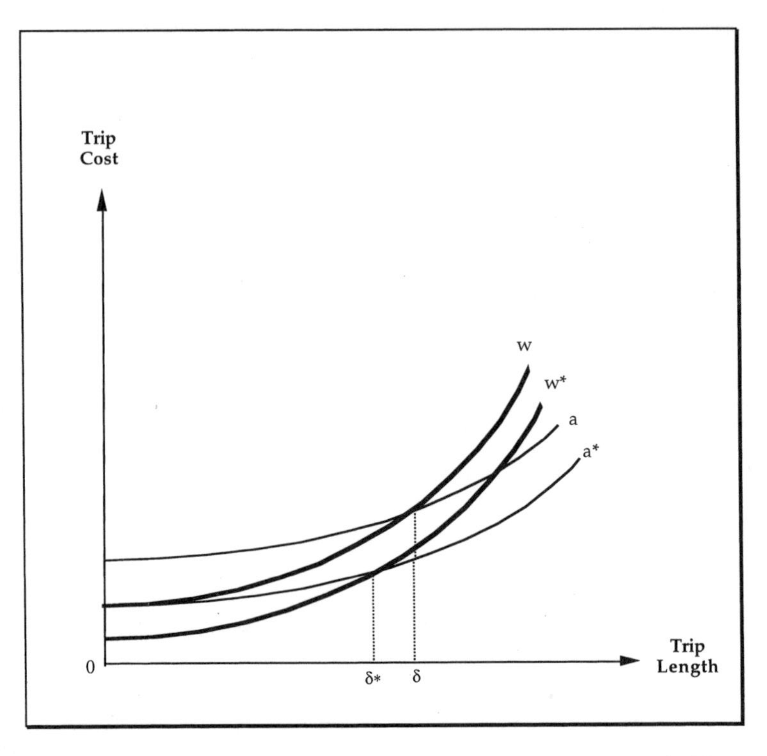

Figure 4.3. The new urbanism backfires. The same comparison as in Figure 4.2, but for an example where per-mile auto costs decrease *more* than walking costs, so maximum walking trip length *falls* from δ to δ*.

fined here as a hedonic index of the quantity and kinds of goods one obtains during each sort of trip. We ignore nontime constraints to emphasize the restriction imposed by the time required for a trip in each mode on the choice of the number of trips in each mode. (These simplifications substantially streamline the exposition while not affecting the qualitative results.)

In this case, the decision process behind the choice of the number of trips may be formalized in a rational choice framework, for example, as the constrained maximization problem of choosing the number of trips by each mode to maximize travel benefits, subject to a budget constraint reflecting travel costs and available time. In the standard functional notation of this modeling approach, the problem statement is to assume that individuals choose their desired number of trips by each mode to maximize $U(\mathbf{a}, \mathbf{w}, \mathbf{b}, x)$ subject to $y = x + \mathbf{ap}_a + \mathbf{wp}_w + \mathbf{bp}_b$, where U is an index of the benefits of using time for each purpose, **a** is a vector of the number of trips by automobile for each purpose, **w** is a vector of the number of trips by walking for each purpose, **b** is a vector of the number of trips by bus or other transit for each purpose, x is a composite of the time spent on other activities, the $\mathbf{p}_i$ are the respective vectors of times for each trip type in each mode ($i = a, w, b$), and y is total available time.

For example, say there are ten different trip purposes that we index by $j = 1, 2, 3, \ldots, 10$. Perhaps the first purpose is grocery shopping, the second trips to school, and so on. Then $\mathbf{a} \equiv (a^1, a^2, a^3, \ldots, a^{10})$ = (*the number of car trips to grocery store, number of car trips to school, . . .*). The total number of car trips taken for all purposes during the reference time period (a week, say) is $\Sigma_{j=1}^{10} a^j$, and the total time spent traveling by car was then $\mathbf{ap}_a = \Sigma_{j=1}^{10} a^j p_a^j$. Note further the time per trip is the quotient of trip length m_i and speed t_i; that is, where $p_i \equiv m_i/ti_i$ for i = a, w, b, for any particular trip purpose (i.e., with superscripts suppressed for simplicity).[5]

The solution to the choice problem is summarized by the trip demand functions $\mathbf{a}(\mathbf{p}_a, \mathbf{p}_w, \mathbf{p}_b, y)$, $\mathbf{w}(\mathbf{p}_a, \mathbf{p}_w, \mathbf{p}_b, y)$ and $\mathbf{b}(\mathbf{p}_a, \mathbf{p}_w, \mathbf{p}_b, y)$. These functions have many useful properties, but their practical value for the problem at hand is that for any given set of travel preferences, they transparently relate changes in trip costs to the number of trips desired, by trip purpose. For example, they can be used to estimate the impact of an urban design change that lowered the time (or other cost) of a trip by foot on the number of trips by foot, the number of trips by car, and the number of transit trips—for each trip purpose. This information could thus be used to calculate how vehicle miles traveled (VMT) respond to increased pedestrian, transit, or auto access due to a change in street patterns. Estimable forms of these demand functions for empirical application to specific data may be obtained by specifying a particular form for U (e.g., Small, 1992).

However, the basic theoretical implications of the behavioral model can be explored without data. At least one potential inconsistency regarding the transportation benefits of alternative land-use patterns at the margin is *internal* to the design principles. To show this most clearly, this chapter is restricted to deriving some basic implications of the behavioral model via the method of comparative statics.[6]

In the context of the model presented above, how can the pivotal features of the new plans be represented? Rather than attempt a comprehensive review, the analysis is restricted to the three most common design elements with assumed transportation benefits: a gridlike street pattern intended to reduce the distance between local trip origins and destinations, "traffic-calming" measures intended to slow cars down, and integrated land uses at higher densities intended to combine more trip destinations at single locations.

The role of the grid in these plans is multifaceted, ranging from increasing the "legibility" of the neighborhood to improving the connection of people and places. Among the ideas that have been reborn in the New Urbanism, the renewed popularity of the grid is both the most frequently mentioned by traffic analysts and perhaps the most compatible with modern street and subdivision codes. For transportation purposes, its major function seems to shorten local trips.

The relationship between the time required for each trip in each mode and land use is assumed to be captured by way of a "grid shift" parameter γ, where an increase in γ (more gridlike design) decreases trip lengths. That is, the derivative $dm_i/d\gamma < 0$ for i = a, w, b. Notice this parameter could represent the effect of any land-use change that made a trip shorter. It is also compatible with a specification assuming that transit or pedestrian trips are shortened more than are car trips, where $dm_a/d\gamma > dm_w/d\gamma$, or other possibilities.

Traffic calming refers to the narrowing of streets and intersections, and other means as well, that slow cars down (e.g., Untermann, 1984; Ben-Joseph, 1997a, 1997b). We model this effect with a "calming" parameter χ, where an increase in χ slows car speeds down; that is, $dt_a/d\chi < 0$. Finally, mixing, combining, and intensifying the density of land uses to make any one trip potentially serve more than one purpose might affect trip demand in at least two ways: It can essentially increase the consumption associated with a trip directly and it can also lower the cost of any "chained" trip. Defining an increase in the shift parameter μ to symbolize an increase in land-use mixing or more intense use of a destination site, the former effect can be represented by $\partial a/\partial\mu < 0$ and the latter by $dp_i/d\mu < 0$, again for i = a, w, b. More intensive use can also increase congestion, such that $dt_a/d\mu < 0$.

The Generic Impacts of Design Features on Travel

With these design features so defined, we can investigate their individual and collective theoretical impacts on travel behavior via comparative statics, as in the discussion below.

The Circulation Pattern

We want to examine how travel behavior is, in theory, influenced by the street circulation plan. Our approach is to see how things change in the model set up above when the circulation pattern is changed in a way that shortens trips. Or, in our notation, what happens to VMT when γ rises?

Note first that total VMT for all car trips is $\mathbf{am}_a = \Sigma_{j=1}^{10} a^j m_a^j$. Hence, an approximate measure of the effect on VMT for one particular purpose due to a move toward a grid street pattern is simply $d\text{VMT}/d\gamma$, where

$$\frac{d\,\text{VMT}}{d\gamma} = a\frac{dm_a}{d\gamma} + m\frac{da}{d\gamma}. \tag{1}$$

(This approach treats trips as approximately continuous. But of course they are not, and the modifications necessary to account for the discrete nature of the trip decision are described in Ben-Akiva and Lerman [1985], Train [1986], and Small, [1992].) Equation (1) succinctly summarizes the automobile travel behavior of an individual benefiting from a more gridlike street network that in turn leads to shorter trips. The first term on the right side of equation (1) measures the effect of shorter trips for the number of car trips prior to the street network change. It enters equation (1) negatively by assumption, and summarizes the results of the studies that have held trip frequency fixed. The second term on the right side of equation (1) is the induced effect on the number of car trips. Do we expect trips to increase or fall?

To see this, note that the number of car trips responds to a small change in trip length according to the total derivative

$$\frac{da}{d\gamma} = \frac{\partial a}{\partial m_a}\frac{dm_a}{d\gamma}\frac{1}{t_a} + \frac{\partial a}{\partial m_w}\frac{dm_w}{d\gamma}\frac{1}{t_w} + \frac{\partial a}{\partial m_b}\frac{dm_b}{d\gamma}\frac{1}{t_b}. \quad (2)$$

The first term on the right side is the change in the desired number of car trips induced by the time savings per trip. This is likely positive, as can be seen from the Slutksy decomposition for $\partial a/\partial p_a$, which breaks down the price change impact into two parts: the impact due to a change in relative prices, and the impact due to a change in overall costs:

$$\frac{\partial a}{\partial p_a} = \frac{\partial a^C}{\partial p_a} - a\frac{\partial a}{\partial y},$$

where $\partial a^c/\partial p_a \equiv \partial a(p_a, p_w, U)/\partial p_a < 0$ is the change in demand due to the change in relative prices (the "compensated" effect) and $\partial a/\partial y$ is the impact of having time freed up by the shorter trips. If automobile trips are a normal good (i.e., the demand for auto trips increases with resources), then $\partial a/\partial y > 0$ and $\partial a/\partial p_a$ must be negative.

Thus, the demand curve for automobile trips is typically downward sloping, as expected, and the first term in equation (2) is positive: All things considered as the time per trip falls, due in this case to a shorter trip, people will tend to want to take more trips.

The number of car trips can fall with a decrease in trip length, however, if the sum of the second and third terms in equation (2) is sufficiently negative. These represent the cross-price or substitution effects of shorter walking and transit trips on car trips. As walking trip lengths fall, owing to a better system of walkways, more direct street patterns, and so on, we might expect people to substitute walking trips for car trips. Indeed, pedestrian trips are more influenced by trip length (and purpose) than by trip time, especially when compared to motorized transport. Evidence that walking trips fall off dramatically after trip distance of a half-mile (e.g., Untermann, 1984) suggests that the second term in (2) is highly elastic near that figure, and zero for longer distances. Shorter transit trips have a less clear effect, again depending on the trip purpose and other particulars not explicitly modeled here—though the time of the trip is probably a more important single indicator than the trip length.

Hence, if automobile trips are a normal good, then $\partial a/\partial m_w$ is negative and the sign of equation (2) is indeterminate. If the new street network is such that people tend to substitute walking or transit for car trips compared to an alternative plan, and the demand for car trips is relatively insensitive to the length of the trip, the number of car trips can fall. But if these conditions are not met, car trips can rise. Whether VMT rises or falls is a separate matter. VMT is the product of the number of trips and their length. If trip lengths fall, as implied by a move to a grid, (2) shows that the number of trips could rise—especially where few transit or walking trips are substituted for car trips and if car trips are sensitive to their length. If the number of car trips rises enough, then VMT could rise as well.

To see this, look at how VMT for a given trip purpose changes with an increase in the grid parameter:

$$\begin{aligned}\frac{d\mathrm{VMT}}{d\gamma} &= a\frac{dm_a}{d\gamma} + m_a\frac{da}{d\gamma} \\ &= \left(1+\varepsilon_{ap_a}\right)a\frac{dm_a}{d\gamma} + m_a\left(\frac{\partial a}{\partial m_w}\frac{dm_w}{d\gamma}\frac{1}{t_w} + \frac{\partial a}{\partial m_b}\frac{dm_b}{d\gamma}\frac{1}{t_b}\right),\end{aligned} \tag{3}$$

where $\varepsilon_{ap_a} < 0$ is the own-price elasticity of demand for trips by car. A sufficient condition for the right side of equation (3) to be negative, and hence for VMT to be lower in more gridlike neighborhoods, is that trip demand be price inelastic (i.e., $\varepsilon_{ap_a} > -1$) and the cross-price elasticities be negative. In that case, the number of desired car trips does not increase enough to offset the shorter trip distances, and total travel falls. (This is more likely the slower the trip.) If the price elasticity of trip demand is elastic and the cross-price elasticities are sufficiently small, however, VMT will rise.[7]

More simply, a move to a grid shortens trip lengths for all modes. The demand for trips in each mode will then likely rise. In part, however, this depends on how well one mode substitutes for another for a given trip purpose and how the resulting trip lengths suggest for the feasibility of either walking or transit. Even with more car trips, VMT may fall—or it may rise.

Traffic Calming

The remaining results can be obtained with much less work. The effect of slowing car speeds can be assumed unambiguously to lower the demand for car trips. That is, $da/d\chi < 0$ and VMT must fall:

$$\frac{d\mathrm{VMT}}{d\chi} = m_a\frac{\partial a}{\partial t_a}\frac{dt_a}{d\chi} < 0$$

While this feature is an important part of many new plans (e.g., Seaside, Florida), it is also among the most difficult to put into practice. Lower capacity streets and narrower intersections conflict with most transportation and subdivision trends and standards (see, e.g., Reps, 1965; Kaplan, 1990; Bookout, 1992a; Southworth and Ben-Joseph, 1997; Ben-Joseph, 1997a, 1997b).

Mixed and Intensified Uses

These design elements refer to practices that try to encourage residents to accomplish more with each trip, perhaps by bundling more trip destinations at a given node, apart from reducing trip lengths or slowing traffic. Many mixed-use strategies effectively do all three, but in this section we want to isolate the impacts of these plans that are different from those discussed above. Afterward we will consider their cumulative effect.

As discussed above, mixing and intensifying uses has two clear consequences for the travel environment: It essentially increases the potential yield of any one trip, and it reduces the effective cost of additional trips. In the first view, a given trip can accomplish more. Therefore, you do not need to travel as often to obtain a given set of goods. An increase in the mixed-use parameter thus reduces the demand for car trips: $\partial a/\partial \mu < 0$. In the second view, the marginal cost of all trips beyond the first are lower if they can be "piggy-backed" onto the first. This effect on car trip demand is positive: $(\partial a/\partial p_a)(\partial p_a/\partial \mu) > 0$. These two effects overlap somewhat, but both seem to capture part of what would happen, and the net influence is again indeterminate, as

$$\frac{da}{d\mu} = \frac{\partial a}{\partial \mu} + \frac{\partial a}{\partial p_a}\frac{\partial p_a}{\partial \mu} \gtrless 0.$$

A third potential effect is that higher densities could increase congestion, thus increasing trip times. Wachs (1993a, 1993b), for example, has pointed out that while the per-capita VMT is lower in such densely developed and populated places as New York, Hong Kong, and Singapore, congestion is climbing and VMT per square mile is very high. Congestion in turn might depress the demand for car trips relative to walking and transit, depending on how well transit fared with the new densities.

One could argue that the first factor dominates the second; that is, since a given quantity of goods can be obtained with fewer trips, the stimulative impact of the lower cost of chained trips is only secondary. That seems likely in many situations, but it is not axiomatic. The impact of the third potential effect is impossible to generalize without more structure and detail, but congestion may well reduce the number of car trips demanded. Again, the net effect on trip frequency and mode choice is uncertain. The effect on VMT is also unclear, in part because there is the added possibility that walking trips would substitute for car trips—but this seems unlikely for most trip purposes, especially where goods are to be carried back home.

Table 4.1 summarizes these individual results. A move toward a street grid increases the number of car trips demanded, with an uncertain net affect on VMT. Traffic-calming measures reduce car trips and VMT. Mixing and intensifying uses probably reduces trip demand and thus VMT, but it may not, depending on the manner in which it is implemented, the congestion induced, and the purpose of the trips. Table 4.1 also lists the effects of each element on automobile mode split, and the

Table 4.1: Qualitative Effects of Different Neighborhood Design Features on Car Travel

	Design Element			
Traffic Measure	Grid (shorter trips)	Traffic Calming (slower trips)	Mixing Uses and Land-Use Intensification	All Three
Car trips	Increase	Decrease	Increase or decrease (depends on trip purpose and length, and induced congestion)	Increase or decrease (depends on relative mix of elements)
VMT	Increase or decrease (depends on sensitivity to trip length by trips of each mode)	Decrease	Increase or decrease	Increase or decrease
Car mode split	Increase or decrease	Decrease	Increase or decrease	Increase or decrease

cumulative effects of all three features on each behavior: While the details of any one plan would provide a more precise outcome, in general a combination of these features may either increase or decrease both automobile trips and VMT.

However well intentioned, land-use changes intended to reduce car traffic can actually cause problems when naively applied. A second purpose of this chapter is to suggest how such problems can be avoided. The next section shows how the behavioral framework can be used as the basis for comparing the impacts of different plan elements on traffic and pedestrian travel in practice. By way of example, chapter 5 then applies this approach to actual data.

Issues for Applied Empirical Work

The framework in the preceding sections illuminates several issues important for empirical research. Each is discussed below.

Urban Design Influences the Cost of Travel

This is the fundamental insight of the behavioral framework. Those urban design strategies aimed at changing either the time cost of traveling on different modes, the "psychic" cost of travel, or both, can be analyzed within a consumer demand framework much as one would analyze price changes for other goods.

While this idea is straightforward and provides a ready link to a large literature on how consumption changes with prices, the architects and planners popularizing the new urban designs have rarely spoken in these terms. Street grids, mixed land uses, and inviting pedestrian neighborhoods all are intended to change either the time cost of traveling (e.g., by placing origins and destinations in more direct proximity) or the relative cost across modes (e.g., by slowing auto travel and facilitating nonautomobile alternatives). Similarly, the more aesthetically oriented design elements, such as plazas and streetscapes, are intended to alter what we will call the relative "psychic" cost of travel on different modes, for example, by making walking trips more pleasing.

Urban Design and Nonwork Travel

Many of the new urban designs are implicitly and sometimes explicitly aimed at influencing nonwork travel. The goal is often to cluster shopping, entertainment, or other nonwork destinations closer to residences, and to encourage people to walk or use transit to get to those destinations. While work trips are also mentioned (most notably in transit-oriented development plans proposing to facilitate transit commuting), the more discretionary nonwork trips play a prominent role in the transportation plans of modern-day urban designers.

At an intuitive level, this focus on nonwork travel is appropriate. Persons have more discretion over nonwork travel, and so might presumably be influenced more by the proposed land-use and urban design changes.[8] Furthermore, close to 80 percent of all urban trips are for purposes other than commuting to work, so there is much to be said for increasing the focus on nonwork travel (NPTS, 1993). Yet an emphasis on nonwork travel brings with it the disadvantage that we know very little about the determinants of nonwork travel behavior.

Transportation demand models are still based on the four-step method, a model of commuting flows, developed in the 1950s and 1960s (see chapter 1). The behavioral framework of the model is quite weak, even for the work trip that it purports to predict, but it worked tolerably well when restricted to commuting behavior. The assumptions and performance of the four-step model break down severely, however, when applied to nonwork travel.

Transportation engineers and economists have for years been trying to develop alternatives to the four-step model of travel demand estimation. This includes attempts to develop travel demand models with a solid behavioral foundation that can predict both commuting and nonwork travel. The complexity of that task has proven enormous. While progress has been made, there is still no commonly accepted model that reliably predicts nonwork travel with data and computing requirements modest enough to allow the technique to move out of a research environment. The implication is that the designers and planners who propose linking urban form and transportation policy have, possibly unwittingly, bumped into one of the frontier topics of travel behavior research. Thus, thinking about nonwork travel and urban design is at least as much a research agenda as it is a policy program.

The Complex Choices That Influence Nonwork Travel

If urban design can influence the price of travel, so can many other things. Importantly, travelers themselves can influence the cost of trips through decisions about where to live, where to travel, and when (e.g., what time of day) to travel. This makes travel demand more complicated than many other types of consumer behavior.

Transportation economists have long tried to deal with that complexity by incorporating a range of choices within their models. The discrete choice models pioneered by McFadden and others (McFadden, 1974; Domencich and McFadden, 1975) model travel demand as the outcome of a several linked choices, potentially including choices about when to travel, how to travel (i.e., by which mode), and where to travel. Activity models have a similar motivation, although they use different modeling techniques (e.g., Kitamura and Recker, 1985; McNally, 1997). As mentioned above, this approach is complex and has not yet produced a model that can be implemented by practicing planners and engineers.

Our approach attempts to strike a balance between the complexity of activity-based and discrete-choice frameworks and the limited behavioral foundations of the traditional four-step model. We strike that balance in the following way. First, we base our empirical tests on the behavioral framework described above, thereby placing our work squarely in the realm of theories of consumer behavior. Second, we focus on empirical tests of hypotheses, rather than on predicting nonwork travel patterns. This simplifies our task. Our goal is not to reliably predict nonwork travel flows. Instead, we propose the more modest goal of reliably testing whether there is a link between nonwork travel and urban design.[9] Third, while not incorporating all possible traveler choices into the model (that, we believe, would swamp us with unnecessary complexity), we incorporate the most important choice margins into our empirical analysis. That gives our analysis more credibility and suggests how future research might similarly incorporate other choices into more complex empirical models.

Trip Generation and Mode Split

The theoretical model in the Behavioral Framework section above confounds the choice of the number of trips and the mode of travel. In terms of the four-step model, this confounds trip generation and mode choice. For simplicity, we separate the trip generation and mode choice components of the framework in the empirical implementation.

Land Use Near Origins and Destinations

Land use near both origins and destinations of trips can potentially influence travel behavior. For several trips linked together in a tour (called "trip chaining"), land use throughout the travel route can be important. This has been recognized in the recent design literature (e.g., Bernick and Cervero, 1997), but most data sources are limited in their geographic detail. The travel diary data used in chapter 5, like most such data, have no information on the exact location of nonwork trip destinations. For that reason, we focus on land-use patterns near travelers' residences. While this is a necessary simplification given our data restrictions, land use near residences is prominently mentioned in the new urban design literature and is an appropriate place to begin an empirical evaluation.

The Importance of Geographic Scale

The ideas associated with the new urban designs are typically at a neighborhood scale. This is due in part to the emphasis on pedestrian travel, and in part to a concern for neighborhood character and sense of place in this emerging literature. Because of the fine geographic detail emphasized in the new urban designs, most recent research on travel behavior and design has measured land-use characteristics for census tracts or similarly small geographic areas.[10] Yet census tracts are often much smaller than the distance of nonwork trips. Tracts can be only two to three miles across at their widest point in urban areas. Many nonwork trips, especially for the data used in chapter 5, are longer.[11]

For nonwork trips that extend over several miles, land use immediately near any one location (e.g., a trip origin) might be an incomplete description of the land-use characteristics that influence travel. This is related to the need to identify land-use characteristics near both trip origins and destinations. If one can only measure land use near a person's residence (as is often the case), measuring that land use for increasingly larger areas can potentially capture information on land use near some nonwork trip origins. Yet larger geographic areas also obscure the land-use character in the immediate neighborhood of residence. Overall, different levels of geographic detail should be tested in empirical research. With the exception of Handy's (1993) explicit test of the importance of geographic detail in linking land use and travel, this idea has been almost completely overlooked in the recent literature.

The next section addresses many of these issues.

An Empirical Strategy

Our basic approach, informed by the ideas above, is to regress trip behavior variables on variables that include measures of land use. The regressions are derived from the model presented in the Behavioral Framework section of this chapter. The dependent variable for the regressions measures the number of nonwork automobile trips (trip generation).

The data come from two travel diaries for southern California. One surveyed persons in Orange County and nearby parts of greater Los Angeles (hereafter called the Orange County/Los Angeles data), and the other diary is from San Diego.

Travel diaries ask respondents to keep a log of all trips made during a particular time period—usually one or two days. Trips are classified according to purpose (e.g., work, work related, shopping, entertainment, education), allowing us to identify nonwork trips. Unlike many transportation demand models, including those based on the four-step method, travel diary data are available at the individual level. The ability to deal with individual rather than aggregated data allows us to link the empirical model more closely to the individual behavioral framework described above.

The regression model of travel behavior includes three classes of independent variables. First, we include the price of travel and the income level of the individual or household, analogously to other consumer demand models. Second, we include several sociodemographic variables that are known from prior research to influence driving behavior, presumably via their influence on preferences, such as gender, education levels, and age and number of persons in the household (Vickerman, 1972). Third, we include measures of land-use and urban design characteristics near the residences of the individual travel diary respondents.

The model above yields travel demand functions for three different modes—automobile, walking, and transit—as a function of the price of travel on each mode and individual income. The travel demand functions, the number of trips that an individual wants to take on each mode at each price, are

$$\begin{aligned} a &= f(p_a, p_w, p_b, y) \\ w &= f(p_a, p_w, p_b, y) \\ b &= f(p_a, p_w, p_b, y) \end{aligned} \tag{4}$$

where, as before, a, w, and b are the number of automobile, walking, and transit trips respectively; p_a, p_w, and p_b are the trip prices on those modes; and y measures the resources available for travel.

As discussed above, travel is a derived demand based on the demand for activities and goods that require travel. Yet a fully operational derived demand model for travel behavior is more complex than is necessary to get insights into the research questions examined here. The behavioral framework described above is a simplification that allows us to represent travel within the framework of consumer demand theory without requiring that we model the complicated details of individual demand for the various goods and activities that require travel.

We further simplify the framework in equations (4) by focusing on only one mode of travel at a time. This yields trip generation functions for each mode. To be consistent with the recent literature's emphasis on reducing automobile travel, we focus on nonwork automobile trips:

$$a = f(\mathbf{p}, y; \mathbf{S}) \tag{5}$$

where

a = the number of nonwork automobile trips taken by an individual,

$\mathbf{p}$ = the vector of relative time costs (or prices) of a nonwork automobile trip (note that we use only the scalar form p hereafter),

y = individual travel budget, and

$\mathbf{S}$ = a vector of sociodemographic shift (or taste) variables.

The trip generation model in equation (5) is consistent with our data for southern California, where most travel is by private automobile.[12] For urban areas with a greater diversity of travel options, travel on other modes should be examined in more detail.

In general, travel expenses include both money and time. However, our samples are limited to private automobile users who are faced with similar money costs.[13] Hence, the model is simplified to consider only the time cost of travel. The time cost of travel varies across individuals depending on their respective values of time. Differences in individual time value are captured by income and other sociodemographic characteristics. As suggested by Kitamura, Mokhtaria, and Laidet (1997), income squared (y^2) can help control for both the extent to which nonwork trip-making is a normal good and the extent to which time spent driving is more valuable (and thus more costly) for persons with higher income.[14]

The time cost is a generalized time cost from the person's residence to all possible nonwork destinations. Thus, the variable p measures *accessibility,* which can be influenced by densities, street grid orientation, the mix of commercial and residential uses, and other land-use characteristics. This is shown by

$$p = f(\mathbf{L}), \tag{6}$$

where $\mathbf{L}$ is a vector of land use characteristics to be defined below.

At this point, there are three possible choices in modeling technique:

Model 1: Price Variation Completely Determined by Observable Land-Use Characteristics

If the differences in time costs of nonwork trips can be completely explained by differences in land-use patterns, equation (6) can be substituted into equation (5) to yield

$$a = f(\mathbf{L}, y; \mathbf{S}) \tag{7}$$

The model in equation (7) is a reduced form reflecting the assumption that differences in the time cost of travel are due to differences in land use and urban design at different locations. The advantage of the model in equation (7) is that the time cost of travel is potentially endogenous in a trip generation regression (because, e.g., persons choose residential locations or departure times for reasons that are correlated with their desired travel behavior), but land-use and urban design characteristics are more plausibly exogenous to individual travel behavior.[15] The disadvantage of this model is that if land use and design are measured incompletely, there might be differences in the time cost of travel even after the land-use variables are introduced into a trip generation regression. This suggests the next model.

Model 2: Include Both Price and Land Use Variables in the Trip Generation Regression

Perhaps, however, price effects are not completely captured by land-use variables. This is more likely the more mixed the travel patterns, the higher the number of employment centers and other major travel destinations in the region, and the more diverse the commute and nonwork travel options, among other things. Particularly in large metropolitan areas, as studied here, it seems improbable that land uses and urban design completely reflect the costs of each trip taken.

Both the price variable and the land-use variables can be used in a regression equation, as shown by

$$a = f(p, y, \mathbf{L}; \mathbf{S}). \tag{8}$$

The time-cost variable can be broken down into two components: trip distance and trip speed. These variables can be more easily linked to policy:

$$a = f(m, t, y, \mathbf{L}; \mathbf{S}), \tag{9}$$

where m and t denote the median nonwork trip length and speed for each travel diary respondent. Median, rather than mean, trip lengths and speeds were used to reduce the influence of extreme patterns.

While the model in equation (9) controls for price variation beyond what can be measured with the $\mathbf{L}$ variables, trip distances and speeds are potentially endogenous. Without being able to treat median distance and speed as endogenous variables (due to data limitations), we report results from both model 1 and model 2 here, to demonstrate that the effect of the land-use variables does not vary much across the two specifications.[16]

Model 3: A Two-Step Procedure

The possibility that trip costs may be at least partly up to the individual traveler could introduce a form of simultaneity bias in our statistical results if not properly accounted for. Our response to this threat is a two-step procedure that can be implemented by first regressing the price variables (each individual's median nonwork trip distance and median nonwork trip speed) on land-use characteristics near that person's residence, as suggested by equation (6) and shown by

$$m = f(\mathbf{L}), \tag{10}$$

$$t = f(\mathbf{L}).$$

The predicted distances and speeds from that regression will, by construction, be uncorrelated with other determinants of trip prices. Those predicted values can then be used in the trip generation equation shown in equation (5), to yield

$$a = f(\hat{m}, \hat{t}, y; \mathbf{S}), \tag{11}$$

where $\hat{m}$ and $\hat{t}$ are the predicted median nonwork trip distance and speed from the first stage regression in equation (10).

Equation (11) is a reduced form that includes the effect of land use on trip prices (through the effect on median trip distance and median speed) and the effect of prices on trip generation. The disadvantage of this model is that it obscures the relationship between land use and trip generation, since land-use variables do not appear in equation (11). Because that complicates the interpretation, the results of model 3 might not, by themselves, be useful for policy analysis. Yet model 3, when combined with the results of models 1 and 2, provides a more detailed understanding of how land use and design influence the components of trip prices (costs) and thus travel behavior.

Summary

The theory and empirical models developed in this chapter are general and can be applied to data from most any urban area. In chapter 5 we illustrate this point by fitting the empirical models on two data sets from southern California. Of course, the policy questions are most interesting in cities that have emphasized new urban design practices or that have a variety of land-use patterns and neighborhood types. On that count, many locales other than southern California might be logical tests for the empirical framework.

Yet several points make a test using southern California data interesting. First, southern California is not the homogeneous, auto-dominated landscape that many assume. There is diversity in neighborhood types throughout the region. Second, the new urban designs have often been proposed in rapidly growing urban areas such as southern California, and recent evidence suggests that those designs have had much impact on planning thought in Los Angeles, Orange County, and San Diego (Reckhard, 1998). Third, and most important, the empirical examples in chapter 5 are intended to illustrate and test the behavioral framework developed in this chapter, and to provide examples of how similar analyses can be conducted for other urban areas.

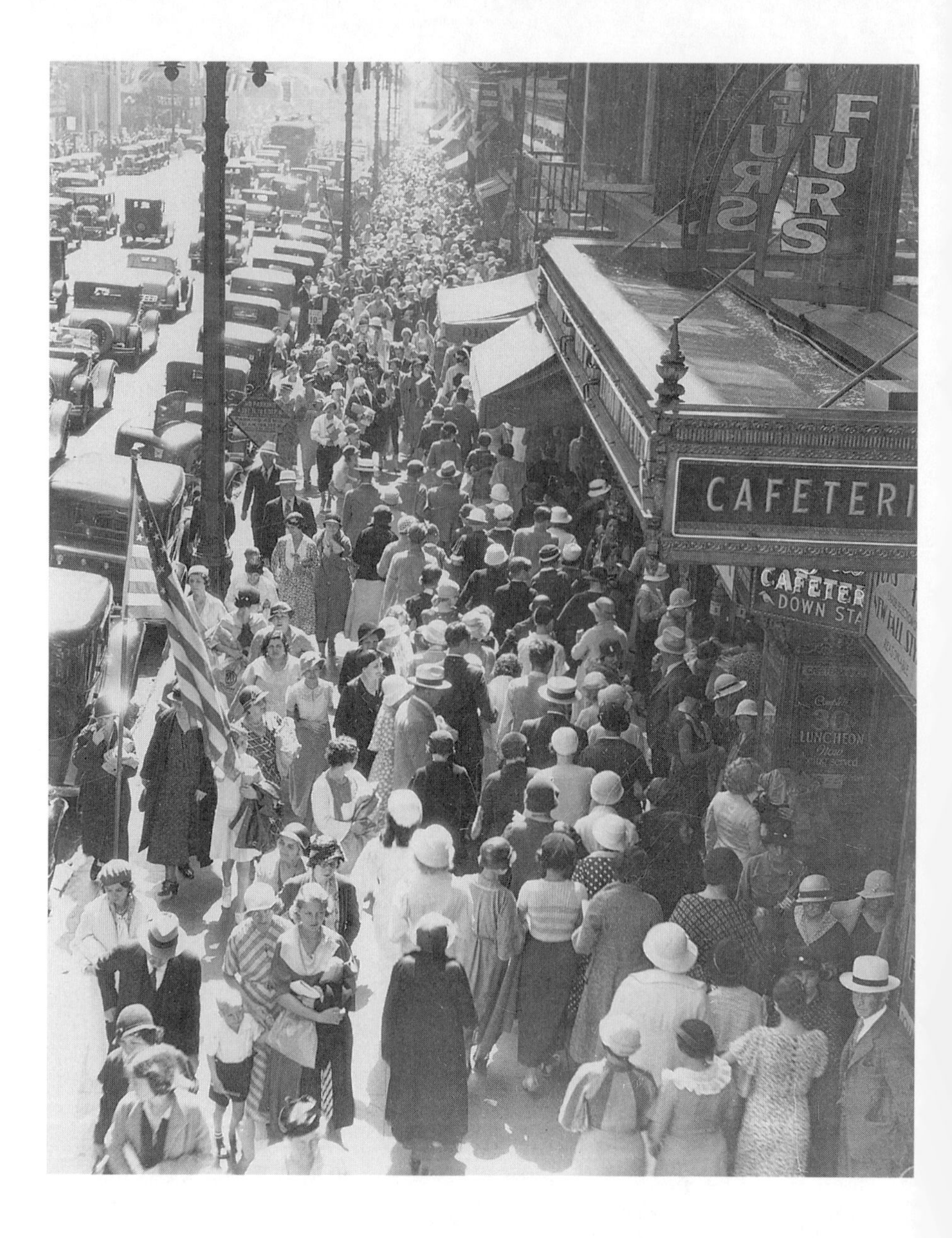

The theory and empirical models developed in chapter 4 are general and can be applied to data from any urban area. In this chapter we illustrate that point by fitting the empirical models on two data sets from southern California. Following a discussion of land-use variable measurement issues and estimation procedures, we show the statistical results to be sensitive to the form of the empirical strategy. We find that, first, land use and design proposals, if they influence travel behavior, do so by changing the price of travel. Second, such impacts become evident in our data only when the endogeneity of residential location choice and issues of geographic scale are incorporated into the regression model.

5 An Empirical Study of Travel Behavior

Overview and Data

The empirical strategy described here involves matching travel diary data to land-use characteristics for two different southern California regions.[1] The first data set is from Orange County and nearby parts of suburban Los Angeles, and the other data are from San Diego. Each data set poses somewhat different opportunities and challenges in measuring the factors of interest, so we describe these in turn.

Orange County/Los Angeles Data

The Orange County/Los Angeles travel diary data set includes data for 769 southern California residents. These were obtained from a 1993 survey administered as part of the Panel Study of Southern California Commuters.[2] Because that survey includes each respondent's street address, we were able to match the travel diary data to land-use variables from the 1990 census and from the Southern California Association of Governments (for the years 1990 and 1994).

The travel diary covered a two-day period, and respondents were preassigned days, so trip making on all days of the week is represented in the data. Individual respondents were first contacted through their employer, and then for follow-up waves the same persons were contacted at home. The sample is employer based, and consequently the respondents are not a random sample of southern California residents. About half of the respondents worked at the Irvine Business Complex, a large, diversified employment center near the Orange County Airport, and the other half worked elsewhere throughout the Los Angeles metropolitan area.

The descriptive statistics shown in table 5.1 illustrate how the survey oversampled Whites, highly educated persons, and persons with high income. This suggests some caution is warranted in interpreting beyond these individuals. Yet restricting attention to an educated, upper-middle-income, largely suburban population still provides interesting information, because many of the new urban designs are intended for low-density, suburban environments with demographics similar to those in this sample.

The dependent variable for the model is the number of nonwork automobile trips made by an individual during the two-day travel diary period.[3] The sociodemographic variables in the model [the vector **S** in equation (5) from chapter 4] are

- FEMALE: a dummy variable equal to one if the respondent is female;
- AGE: the age of the respondent;
- NONWHITE: a dummy variable that equals one if the respondent is Black, Hispanic, or Asian;
- a dummy variable that equals one if the respondent did not graduate from high school;
- a dummy variable that equals one if the respondent is a college graduate;
- the number of children under age 16 in the household;
- the number of cars per licensed drivers in the household;
- the number of workers in the household;

Table 5.1: Summary of Selected Demographic Variables for Orange County/Los Angeles Data

Variables	Frequency	% Share
Gender		
Male	377	50.9
Female	364	49.1
Race		
White	510	86.9
Non-white	73	13.1
Education		
Did not graduate from high school	10	1.3
High school graduate	93	12.2
Some college	234	30.7
Four-year college degree	188	24.6
Some graduate study	234	30.7
Household income		
Less than $15,000	8	0.5
$15,000–$24,999	20	1.1
$25,000–$34,999	47	2.7
$35,000–$44,999	74	6.4
$45,000–$54,999	100	10.1
$55,000–$64,999	64	8.7
$65,000–$74,999	79	10.8
$75,000–$84,999	87	11.9
$85,000–$94,999	73	10.0
$95,000–$119,999	82	11.2
$120,000–$149,999	57	7.8
$150,000 and over	38	4.2

- a dummy variable equal to one if the respondent's commute is longer than the sample median (which is thirteen miles);
- a dummy variable equal to one if the respondent completed the travel diary during a two-day period that contained at least one work day; and
- an interaction term for the long-commute and work-day dummy variables.[4]

Many of these variables have been included in previous studies of the determinants of individual travel behavior (e.g., Vickerman, 1972).

In measuring the land-use variables, the researcher must decide how to operationalize the multifaceted ideas associated with the new urban designs. Most past empirical work on urban design and travel has measured density, land-use mix, and street geometry (see, e.g., Cervero and Kockelman, 1997, for a discussion). The following variables were used to measure those three characteristics:

- POPDEN: population density, in 1990.
- %GRID: the percentage of the street grid within a quarter-mile radius of the person's residence characterized by four-way intersections.[5]
- RETDEN: retail employment divided by land area. This is intended to proxy for the land-use mix (and especially for commercial land uses) near each person's residence.
- SERVDEN: service employment divided by land area. This is also used, in conjunction with RETDEN, to proxy for the land-use mix of neighborhoods.

The models in this chapter used two levels of geographic detail for three of the four land-use variables in the Orange County/Los Angeles model. Consistent with previous studies, we first measured all variables at the level most closely corresponding to the neighborhood. For POPDEN, data were available at the census block group level. For RETDEN and SERVDEN, data for census tracts were used because block group data were not available. The POPDEN, RETDEN, and SERVDEN data for block groups and tracts are for the year 1990. For %GRID, we measured the street geometry within a quarter mile of each person's residence, based on 1994 census TIGER (Topologically Integrated Geographic Encoding and Referencing) files.

Many of the nonwork trips in our sample were long enough that they did not start and end in the same census block group or tract, so we also used a broader scale of geography that involved measuring POPDEN, RETDEN, and SERVDEN for the ZIP codes containing each respondent's residence.[6] Because of the time needed to construct the street geometry variable, %GRID was not measured at any level other than a quarter-mile circle centered on each person's residence.

Chapter 4 concluded that the expected sign for each of the land-use variables is ambiguous, emphasizing the need for an empirical analysis of this topic. Population density, measured by POPDEN, might proxy for closer origins and destinations, which decreases the cost of trips and thus might encourage more trip making. Yet population density might also proxy for congestion, which increases the cost of trips and can discourage travel. If street geometry lowers travel costs by creating more direct routes, the expected sign on %GRID is positive. If, on the other hand, street geometry proxies for narrow streets or other factors that slow automobile travel speeds, there can be a negative relationship between %GRID and the number of nonwork car trips.

If land-use mixing (measured by RETDEN and SERVDEN) reduces the cost of trips by placing them closer to origins, it might induce more driving. Yet if mixed land uses place nonwork destinations close enough to residences to facilitate walking (or other alternatives to driving), or if mixed land uses are correlated with characteristics that slow automobile travel speeds, land use mixing can reduce the number of car trips even when the total number of trips for all modes increases.

San Diego Data

The travel data for San Diego are from the 1986 Travel Behavior Surveys developed jointly by the San Diego Association of Governments and the California Department of Transportation (SANDAG, 1987b). The sample was obtained using a random telephone number "plus one" method, said to eliminate biases against households that have unlisted numbers and households that have recently moved. Participants answered the socioeconomic questions over the telephone. Subsequently, the travel diary was sent to their home address and the information was collected for their travel characteristics on the designated travel day for that household. In total 2,754 households participated, yielding data for 7,469 persons and 32,648 trips.

For this analysis, the data are restricted to those households where a successful address match was obtained based on the household's phone number (Drummond, 1995, discusses this procedure). The resulting sample yielded 4,199 home-based nonwork trips, summarized in table 5.2 and illustrated by location in the map of San Diego County in figure 5.1. Note that by far the two most common mode choices were automobile and walking, with the latter at about 9 percent. Only about 3 percent of nonwork trips were by transit in this sample.

The dependent variable for the trip generation models described in the following sections is the number of nonwork automobile trips taken by

Table 5.2: Sample Mode Distribution for San Diego Data

Mode	Number of Trips	% of Trips
Auto	3,609	85.95
Walk	369	8.79
Other	221	5.26
Total	4,199	100.0

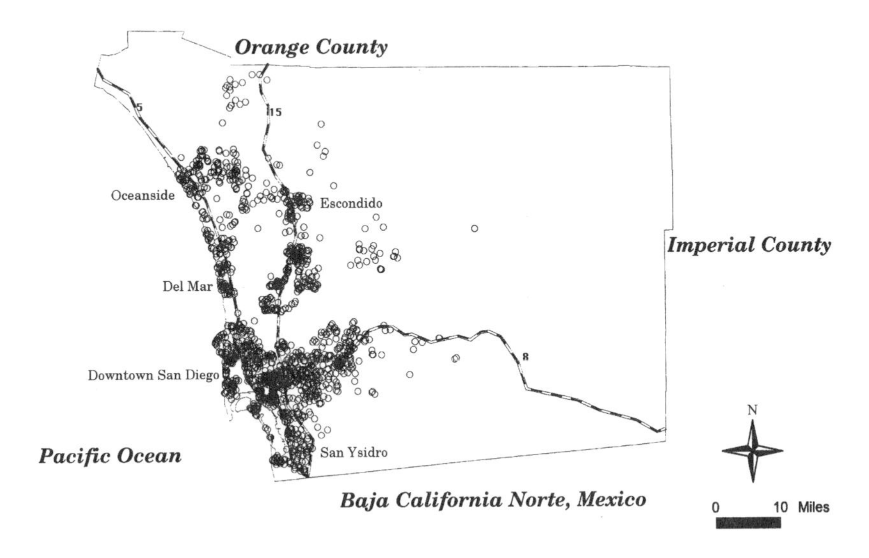

Figure 5.1. Location of sample households in San Diego County.

all individuals in the household during the day covered by the travel diary. Because the San Diego residents were asked to describe household travel, it was not possible to isolate individual travel as with the Orange County/Los Angeles data.

The San Diego respondents were asked to specify their income according to categories.[7] The two income variables included in the model are a dummy variable equal to one if the household had income less than $20,000 per year and a dummy variable equal to one if household income was between $20,000 and $40,000 per year. The sociodemographic (or taste) variables in the San Diego model are

- the mean age of household members;
- the number of children under age 16 in the household;
- HHSZ, household size;
- TENURE, housing tenure (=1 if owner occupied); and
- SFR, type of housing unit (=1 if a detached single-family dwelling).

The land use variables for San Diego are[8]

- Street network variables, GRID and MIXED: The variables describing the street network of the neighborhood are based on a visual inspection of the network within a half-mile of the household, using Geographic Information System software. The network was judged to be either a "connected street" network (GRID = 1), a "cul-de-sac" network (the omitted category in the regressions), or a mixture of the two (MIXED = 1). An example of an observation in a "connected" neighborhood is given in figure 5.2, while

Figure 5.2. Example of a Connected Neighborhood.

the distribution of the sample trips by street pattern is summarized in tables 5.2 and 5.3. As table 5.3 illustrates, the great majority of trips were by car, and most originated in cul-de-sac neighborhoods, though some walking trips and many connected areas are also represented.

Table 5.3: Trip Characteristics by Residential Street Configuration at Trip Origin (San Diego data)

Street Pattern	Duration (in minutes)	Distance (in miles)	Speed (miles per hour)	Number of Trips	Percentage of Trips
Grid	14.88	4.71	17.06	664	15.8
Cul-de-sac	14.59	5.72	22.96	2,010	47.9
Mixed	13.10	4.55	20.19	1,525	36.3
Total	14.09	5.13	21.02	4,199	100.0

- Land-use mix variables, %RESID, %COMM, and %VACANT: The land-use mix variables are the proportion of land in the household's census tract in residential use (%RESID), commercial use (%COMM), or vacant (%VACANT). The land-use mix data were available by census tract for San Diego County for 1986 (SANDAG, 1987a, 1987c). These data were obtained by aerial photography and site visits and thus represent actual land use rather than functional representations based strictly on political or zoning decisions.[9]
- Proxies for density and neighborhood character, D_CBD and D_CBD2: We included the distance of the trip origin from the central business district (D_CBD) as well as that same distance squared (D_CBD2) to account for nonlinearity in the effect of distance. The distance variables are meant to proxy for neighborhood age and density, assuming that newer and more suburban neighborhoods are more distant from the central business district (CBD). (In this case, the distance from the CBD is measured from the intersection of 4th Avenue and B Street in downtown San Diego, per the practice of the California Department of Transportation).
- Street network density, HEAVY: The remaining street network variable identifies dense street patterns (HEAVY = 1). This variable was measured by visual inspection of street maps within a quarter mile of each respondent's residence tract.

The hypothesized signs of the land-use variables for San Diego, like those for Orange County/Los Angeles, are generally ambiguous. The street network variables can proxy either for more direct access (lower trip cost) or for slower travel speeds. A similar point applies to the land-use variables. If the distance variables proxy density, then the effect is theoretically ambiguous. Similarly, heavy street densities might be associated either with more direct travel routes (lower trip costs) or with narrower and more heavily traveled streets, which can be associated with slower trip speeds.

Base Models

The results of fitting the regression models from chapter 4 on the Orange County/Los Angeles data are shown in table 5.4. In column A, we only include the sociodemographic control variables from the model in equation (7) of chapter 4 (the model with only land-use and sociodemographic control variables). Note that the sociodemographic variables generally perform as expected. Based on column A, women make more nonwork automobile trips, older persons make fewer nonwork car trips, nonwork auto trips increase with income (at the 10 percent two-tailed significance level), persons with more children in the household take more nonwork car trips, and nonwork trip making increases with the number of workers in the household (at the 10 percent two-tailed signif-

Table 5.4: Ordered Probit Regressions for Number of Nonwork Automobile Trips (Orange County/Los Angeles data)

	Column A: Demographics Only		Column B: Block Group/Tract Land-Use Variables		Column C: With Median Speed and Distance	
Independent Variables	Coefficient	*Z* statistic	Coefficient	*Z* statistic	Coefficient	*Z* statistic
Trip time-cost variables						
Median nonwork car trip speed					0.0128*	2.12
Median nonwork car trip distance					−0.0183*	−3.24
Land-use variables						
%GRID			0.0024	0.98	0.0025	0.87
POPDEN (1,000 persons/sq mile)			−0.0333	−1.09	−0.0280	−0.76
RETDEN (1,000 retail jobs/sq mile)			0.4218	1.56	0.1086	0.32
SERVDEN (1,000 service jobs/sq mile)			−0.1607	−1.41	0.0018	0.01
Income variables						
Household income ($1,000s)	0.0117	1.89	0.0021*	2.88	0.0122	1.35
Household income squared	−0.0001	−1.60	−0.0001*	−2.41	−0.0001	−1.20
Sociodemographic variables and controls						
Female	0.2979*	2.95	0.4397*	3.70	0.5502*	3.92
Age	−0.0107*	−2.32	−0.0081	−1.54	−0.0093	−1.50
Non-White	−0.1096	−0.75	0.1125	0.66	0.1677	0.85
No high school	−0.6369	−1.07	−0.6790	−1.13	−1.4663	−1.38
College	0.1018	0.97	0.1193	0.97	0.0559	0.38
Number of children <16 years	0.1980*	3.06	0.1886*	2.52	0.1258	1.50
Cars per drivers in household	0.1003	0.84	0.0127	0.09	-0.0751	−0.45
Number of workers in household	0.1228	1.92	0.1008	1.44	0.1300	1.62
Long commute	0.0211	0.11	−0.2380	−1.08	0.1806	0.69
Work day	−0.2744	−1.59	−0.4957*	−2.48	−0.5802*	−2.47
(Long commute) * (Work day)	−0.1075	−0.48	0.1219	0.48	−0.3385	−1.12
N	463		353		259	
log(*L*)	−1027.91		−777.35		−572.83	

*Coefficients significantly different from zero at the 5 percent level.

icance level).[10] These results generally agree both with intuition and with previous research on the determinants of individual trip generation (e.g., Vickerman, 1972; Gordon, Kumar, and Richardson, 1989c; Rosenbloom, 1993).

The land-use variables are added to the regression in column B of table 5.4. The land-use variables in table 5.4 are measured at the census block group level for population density (POPDEN), for quarter-mile radius circles centered on each residence for %GRID, and at the census tract level for retail and service employment densities (RETDEN and SERVDEN, respectively). Thus, the geographic scale in table 5.4 corresponds to the neighborhood level emphasized in most recent research on the effect of urban design on travel behavior. None of the four land-use variables for the Orange County/Los Angeles data are statistically significant, either individually or jointly, in the model in column B.

The price variables, median nonwork car trip speed and distance, are added to the model in column C, which corresponds to the regression model in equation (9) from chapter 4.[11] The price variables are significant with the expected signs: persons with higher median trip speeds make more nonwork automobile trips, and as median distance increases nonwork trip frequency drops. The land-use variables are again insignificant, both individually and jointly, in column C of table 5.4. The results shown in table 5.4 are not meaningfully different when the land-use variables are measured for ZIP codes, so those results are not reported here.

In table 5.5 we report the results from fitting the same models on the San Diego data. Column A of table 5.5 explains trip frequency using only the sociodemographic variables. (Recall that trip frequencies for the San Diego data are for households, not individuals.) The signs on the income dummy variables suggest the same quadratic income effect that was evident in table 5.4. As income increases, household trip frequencies first increase (when household income moves into the $20,000 to $40,000 per year range) and then decrease (for incomes greater than $40,000.) Trip frequency also increases with the average age of the household, with the number of children under age 16 in the household and with household size.[12] The effect of age on trip frequency is opposite from the effect in the individual data reported in table 5.4, and the effect of housing tenure is difficult to interpret, but generally the influence of the sociodemographic variables is similar to the results in table 5.4.

We add land-use variables to the model in column B of table 5.5. The coefficients on GRID and MIXED are significantly positive at the 5 percent level. If gridded street patterns reduce the cost of nonwork automobile trips, this result is consistent with the theory presented in chapter 4. Yet at this point in the analysis the primary message is that the effect of land-use characteristics on trip generation is complicated and potentially contrary to the expectations associated with the new urban designs. The coefficient on HEAVY is significantly negative at the 10 percent level, suggesting a tendency for dense street networks to be associated

Table 5.5: Ordered Probit Regressions for Number of Non-Work Automobile Trips (San Diego data)

Independent Variables	Column A Demographics Only		Column B Land-Use Variables		Column C With Median Speed and Distance	
	Coefficient	*Z* statistic	Coefficient	*Z* statistic	Coefficient	*Z* statistic
Trip time-cost variables						
Median nonwork car trip speed					0.0475*	16.18
Median nonwork car trip distance					−0.0355*	−6.17
Land use variables						
GRID (=1 for grid street network)			0.1874*	1.97	0.1819	1.89
MIXED (=1 for mixed street network)			0.1663*	2.49	0.1432*	2.12
HEAVY (=1 for dense street network)			−0.1101	−1.65	−0.1219	−1.80
%RESID (percent residential)			−0.0420	−0.23	0.0058	0.03
%COMM (percent commercial)			−0.4342	−1.04	0.3187	0.75
%VACANT (percent vacant)			−0.3387	−1.49	−0.4320	−1.87
D_CBD (distance from cbd)			0.0446*	4.08	0.0303*	2.73
D_CBD2 (distance from cbd, squared)			−0.0009*	−3.50	−0.0007*	−2.57
Income variables						
Household income less than $20,000	−0.1558	−1.87	−0.2107*	−2.40	−0.0998	−1.12
Income between $20,000 and $40,000	0.1633*	2.63	0.06001	0.93	0.1138	1.75
Sociodemographic variables and controls						
Mean age of household members	0.0045*	2.26	0.010357*	4.86	0.0125*	5.79
Number of children <16 years	0.1154*	2.36	0.191778*	3.81	0.2079*	4.08
Number of persons in household	0.3980*	10.52	0.389135*	9.96	0.3794*	9.55
Tenure (=1 if owner occupied residence)	0.2468*	3.55	0.205049*	2.77	0.1893*	2.52
SFR (=1 if detached single family dwelling)	−0.0854	−1.29	−0.08823	−0.27	−0.0958	−1.35
N	1450		1336		1336	
log(*L*)	−2776.99		−2600.81		−2455.16	

CBD = central business district; SFR = single family residence.
*Coefficients significantly different from zero at the 5 percent level.

with fewer nonwork car trips. This tension is more consistent with the rhetoric of the new urban designs, and helps emphasize the complex nature of the travel behavior being studied.

The effect of distance from the central business district is quadratic in column B. Households located farther from the CBD tend to make more nonwork car trips, but the effect reverses signs at a distance of approximately twenty-five miles from the downtown. This likely proxies for the effects of density, congestion, and road network quality on trip-making behavior. It is difficult to interpret more precisely the coefficients on distance and distance squared, but the significant coefficients suggest some possible role for land use and urban form as a determinant of travel behavior.

The price variables, median nonwork car trip speed and distance, are added to the model in column C, and both variables are highly significant with the expected signs. The significance of the land-use variables does not change much when the price variables are added to the model.

In tables 5.6–5.9, we report the results of fitting the two-step routine that is described in chapter 4. Tables 5.6 and 5.7 give results for the Orange County/Los Angeles data, and tables 5.8 and 5.9 show results for San Diego. In tables 5.6 and 5.7, no land-use variable is significant at the 5 percent level in the regression for either median speed or median distance. Not surprisingly, predicted median speed and distance, obtained from the first-stage regressions, are not significant in the second-stage regression in table 5.7. The land use variables for Orange County/Los Angeles appear to be weak predictors of trip prices (speeds and distances), and that could help explain the insignificant coefficient on the land-use variables in the earlier Orange County/Los Angeles models reported in table 5.4.

Table 5.6: Two-Step Method for Nonwork Automobile Trip Frequency (Orange County/Los Angeles data): Step 1, Get Predicted Speed and Distance

	Dependent Variables			
	Median Speed		Median Distance	
Independent Variables	Coefficient	T statistic	Coefficient	T statistic
%GRID	−0.0511	−1.67	0.0150	0.34
POPDEN	−0.0005	−1.36	−0.0001	−0.23
RETDEN	−2.0657	−0.61	−6.5425	−1.39
SERVDEN	−0.2671	−0.25	2.2214	1.38
Constant	26.4034*	22.61	13.4419*	8.14
N	399		454	
R^2	0.02		0.01	
F statistic†	$F(4,394) = 2.33$		$F(4,449) = 0.67$	

*Coefficients significantly different from zero at the 5% level.
†Five percent critical value for both $F(4,394)$ and $F(4,449)$ is 2.37.

Table 5.7: Two-Step Method for Nonwork Automobile Trip Frequency (Orange County/Los Angeles data): Step 2, Use Predicted Speed and Distance in Ordered Probit for Nonwork Car Trip Frequency

Independent Variables	Coefficient	*Z* statistic
Female	0.4399*	3.70
Age	−0.0076	−1.45
Non-White	0.1135	0.67
No high school	−0.7152	−1.20
College	0.1184	0.97
Household income	0.0210*	2.94
Household income squared	−0.0001*	−2.48
Number children <16 years	0.1916*	2.56
Cars per drivers in household	0.0466	0.34
Number of workers in household	0.1068	1.53
Long commute	−0.2251	−1.03
Work day	−0.5065*	−2.55
(Long commute)*(Work day)	0.1300	0.51
Predicted median nonwork car trip speed	−0.0051	−0.19
Predicted median nonwork car trip distance	−0.0442	−1.13
N	353	
$\log(L)$	−778.73	

*Coefficients significantly different from zero at the 5% level.

Table 5.8: Two-Step Method for Nonwork Automobile Trip Frequency (San Diego data): Step 1, Get Predicted Speed and Distance

	Dependent Variables			
	Median Speed		Median Distance	
Independent Variables	Coefficient	T statistic	Coefficient	T statistic
GRID	0.5094	0.42	0.6386	1.13
MIXED	0.7367	0.85	−0.0489	−0.12
HEAVY	−0.0603	−0.07	−0.3959	−0.97
%RESID	−1.3912	−0.59	−0.4024	−0.36
%COMM	−25.3062*	−4.80	−6.904*	−2.77
%VACANT	2.3135	0.79	0.8929	0.65
D_CBD	0.5584*	4.00	0.0864	1.31
D_CBD2	−0.0103*	−3.07	−0.0013	−0.79
Constant	14.6874*	6.54	4.214*	3.97
N	1342		1342	
R^2	0.0756		0.0255	
$F(8,1333)$†	13.63		4.35	

*Coefficients significantly different from zero at the 5% level.
†Five percent critical value for $F(8,1333)$ is 1.94.

Table 5.9: Two-Step Method for Nonwork Automobile Trip Frequency (San Diego data): Step 2, Use Predicted Speed and Distance in Ordered Probit for Nonwork Car Trip Frequency

Independent Variables	Coefficient	*Z* statistic
Income <$20,000	−0.2048*	−2.34
Income between $20,000 and $40,000	0.0552	0.86
Mean age of household members	0.0098*	4.66
Number children <16 years	0.1901*	3.79
Number persons in household	0.3773*	9.73
Tenure (=1 if owner occupied residence)	0.1921*	2.61
SFR (=1 if detached single family dwelling)	−0.106	1.53
Predicted median nonwork car trip speed	0.0471*	2.13
Predicted median nonwork car trip distance	−0.1088	−1.34
N	1336	
log(*L*)	−2610.18	

SFR = single family residence.
*Coefficients significantly different from zero at the 5% level.

The land use variables for San Diego appear to be better measured. This is especially the case for land-use mix, measured by the proportion of land devoted to residential (%RESID), commercial (%COMM), and vacant (%VACANT) uses. Recall that for Orange County/Los Angeles, land-use mix was measured by the density of retail and service employment, a potentially indirect proxy for land-use character. The regressions in tables 5.8 and 5.9 give further evidence that the land-use variables are better measured for San Diego. The proportion of land in commercial uses (%COMM) is significantly negative in both the median speed and median distance regressions. The variables for distance from the CBD are significant, with opposite signs, in the median speed regression. Predicted median nonwork car trip speed is significantly positive in the second-stage regression in table 5.9.

Overall, tables 5.6–5.9 are consistent with the theory in chapter 4. *When land use variables have an impact on nonwork auto trip generation, that impact is through the effect on trip prices (speed and distance). When there is no link between land use and trip prices (possibly because land use has been incompletely measured), the model gives no evidence of a link between land use and trip generation.*

The San Diego results are especially important in clarifying the potentially complicated influence of commercial concentrations near residential locations. The first-stage regressions in table 5.8 suggest that persons living in tracts with more commercial land use have both shorter nonwork trip distances and slower nonwork trip speeds. The net effect on trip cost is ambiguous, providing important perspective on the wealth of ambiguous or weak evidence in the empirical literature to date. The crucial question for land-use policy is how the competing effects of slower speeds and shorter trip distances net out.[13] This emphasizes that researchers and planners should examine how land use and design attributes influence trip costs (speeds and distances), and from there consider how the effect on trip costs influences trip generation and other characteristics of travel behavior.

Incorporating Choices About Residential Location: An Empirical Model

The regression models in the preceding section implicitly assume that causality flows from land use and urban design to travel behavior. That assumption is commonplace in the recent literature; the studies reviewed in chapter 3 very often leap from observed correlations between urban design and travel behavior to the conclusion that design changes can *cause* changes in individual travel. This thinking overlooks the complexity of travel behavior, and risks confounding observed correlations with causal influences. As discussed in chapter 4, individual travel patterns are the result of a large number of decisions about where to live, where to travel, when to travel, and how to travel. In attempting to discern how urban design influences, for example, trip generation (as in the preceding section), it is important to account for other choices that might be wrapped up with the decision to make nonwork automobile trips. We do that here by incorporating residential location choice into the nonwork automobile trip generation model from the preceding section.

It is quite possible that persons choose their residential location based in part on their desired driving patterns. For example, persons who dislike driving might both drive less and choose to live in a high-density, mixed-use neighborhood that supports transportation alternatives other than driving. If that occurs, the regression estimates in the preceding section are confounded by the residential location choices of individuals. Urban design, in this scenario, might not lure would-be automobile commuters out of their cars as much as the new designs might provide residential neighborhoods for persons who already prefer to drive less.[14]

The research question is whether urban design influences how persons wish to travel or whether preferences about driving are rather fixed and persons simply choose to live in neighborhoods that support their desired travel behavior. Phrased that way, the analysis can get complex rather quickly. A simpler approach is to control for the influence of urban design on residential location choice, and then examine any remaining link between land use and travel behavior.

More formally, the econometric problem is that the land-use variables in the **L** vector in equations (7) and (9) in chapter 4 could be correlated with the error term in the same equations, akin to classic endogeneity bias in least-squares models. To illustrate this problem, and a solution, we simplify the model from equation (7) of chapter 4. Assume that the number of nonwork automobile trips is approximately continuous, such that the number of nonwork automobile trips is given by

$$N = \beta_0 + \beta_1' L + \beta_2 y + \beta_3 y^2 + \beta_4' S + u, \quad (1)$$

where the βs are the parameters to be estimated (with β_1' and β_4' vectors), u is the regression error term, and the other variables were defined previously.

If persons choose residential locations (and thus land-use patterns near their residence) based on unobserved preferences correlated with attitudes about driving, the variables in the **L** vector can be correlated with u, the error term in equation (1). If that occurs, the least-squares parameter estimates for (1) will be biased and inconsistent.[15] As in other situations where independent variables are correlated with the regression error term, a solution is to use instrumental variables.[16]

Choosing instruments for **L** requires some consideration of the determinants of land-use patterns near persons' residences. This in turn requires some consideration of residential location choice. A large literature has studied moving and residential location decisions (e.g., Quigley and Weinberg, 1977; Linneman and Graves, 1983; Sarmiento, 1995). A brief summary is that equilibrium residential locations (and thus land-use patterns near individual residential locations) are the result of matches between individuals (the choosers) and residential sites (the choice set).[17] Thus, the residential location of an individual is a function of individual and location characteristics, as shown by

$$\text{ResLoc}_k = f(\mathbf{C}_k, \mathbf{A}_k), \quad (2)$$

where ResLoc_k denotes the residence location chosen by person k, $\mathbf{C}_k$ are k's sociodemographic characteristics, and $\mathbf{A}_k$ are the characteristics of residential locations, including location-specific amenities such as school quality, the demographic composition of the surrounding neighborhood, and the age of the housing stock in the surrounding neighborhood.

The variables in equation (2), because they explain residential location choice, are potential instruments for the **L** variables in equation (1). Of the variables in (2), the individual characteristics in **C** are likely to be the same as the demographic variables in **S**, leaving only neighborhood amenities (**A**) as allowable instruments.[18] We chose four neighborhood amenities as instruments:

- %BLACK: the proportion of the 1990 census block group, census tract, or ZIP code area population that is Black
- %HISPANIC: the proportion of 1990 block group, tract, or ZIP code population that is Hispanic
- HousePre40: the proportion of 1990 block group, tract, or ZIP code housing stock that was built before 1940
- HousePre60: the proportion of 1990 block group, tract, or ZIP code housing stock that was built before 1960

These demographic and housing stock variables are likely to be correlated with the land-use patterns measured by **L**, but because they describe amenities unrelated to transportation, the instruments above are plausibly exogenous to the error term in equation (1).[19] For all instrumental variables regressions reported below, we measured the instruments at the level of geography that most closely corresponded to the geographic detail of the land-use variables. In practice, this meant using instruments measured both at the block-group/tract level and the ZIP code level for the Orange County/Los Angeles data. Because the land-use variables for the San Diego data are measured in most instances at the level of census tracts, the instruments were measured for census tracts in San Diego.

The research question addressed below is whether the results from the preceding section change once instrumental variables estimation is used. Phrased more directly, when we control for the possibility that individuals choose to live in neighborhoods with particular land-use and design characteristics, does a link from urban design to travel behavior remain? If so, that would be evidence that design can influence travel demand and encourage drivers to change their travel behavior.

Incorporating Choices About Residential Location: Results

The results from fitting the instrumental variables model are reported in three steps—for the Orange County/Los Angeles data with land-use variables measured at the census block group and tract level, for the San Diego data with the land-use variables measured at the census tract level, and for the Orange County/Los Angeles data with the land-use variables measured at the level of ZIP codes. For each step, we discuss both the coefficient estimates and the results of diagnostic tests of model validity.

The question, at each step, is not only whether the instrumental variables technique changes our assessment of the influence of land use on trip generation, but also whether diagnostic tests suggest that the modeling technique has promise in controlling for the influence of residential location described above. These questions are of course linked; the results from models with good diagnostics should be treated more seriously.[20]

Step 1: Orange County/Los Angeles (Block Group and Census Tract Land-Use Variables)

The results of fitting the instrumental variables regression on Orange County/Los Angeles data are reported in table 5.10.[21] They suggest that using instrumental variables does *not* affect the basic conclusion from table 5.4; the land-use variables are statistically insignificant in all instrumental variables routines in table 5.10.

The overidentification statistics reported at the bottom of each column test the null hypothesis that the instruments are orthogonal to the regression error, which is required for instrumental variables to give unbiased and consistent estimates.[22] The statistic is distributed χ^2 with degrees of freedom equal to the number of excluded instruments (four in

Table 5.10: Instrumental Variables Regressions for Nonwork Automobile Trip Frequency (Orange County/Los Angeles data): Block Group/Tract Land Use Variables

	Column A: %GRID		Column B: POPDEN		Column C: RETDEN & SERVDEN	
Independent Variables	Coefficient	T statistic	Coefficient	T statistic	Coefficient	T statistic
Land use variables						
%GRID	0.0125	1.241				
POPDEN (1,000 persons/sq mile)			−0.1877	−0.88		
RETDEN (retail jobs/sq mile)					−3.6414	−0.55
SERVDEN (service jobs/sq mile)					3.5458	0.98
Income variables						
Household income ($1,000s)	0.0458*	2.49	0.0369	1.95	0.0618*	2.42
Household income squared	−0.0002	−1.95	−0.0002	−1.60	−0.0003*	−2.03
Sociodemographic variables and controls						
Female	1.0076*	3.21	1.0755*	3.47	0.8400	1.91
Age	−0.0239	−1.74	−0.0245	−1.76	−0.0353	−1.47
Non-white	0.2323	0.51	0.4990	1.07	0.1952	0.34
No high school	−2.1102	−1.32	−1.7292	−1.06	−1.4880	−0.73
College	0.1380	0.43	0.2726	0.84	−0.0475	−0.11
Number of children <16 years	0.4504*	2.27	0.4688*	2.37	0.5206*	2.08
Cars per drivers in household	−0.1253	−0.35	−0.1917	−0.52	−0.0530	−0.12
Number of workers in household	0.2905	1.57	0.2518	1.35	0.3927	1.51
Long commute	−0.3464	−0.59	−0.4983	−0.85	0.3237	0.36
Work day	−1.1719*	−2.21	−1.1606*	−2.20	−0.5671	−0.73
(Long commute) * (Work day)	0.0459	0.07	0.0788	0.12	−0.7301	−0.72
Constant	1.9921	1.58	3.2713*	1.99	0.7380	0.39
N	354		359		358	
Overidentification test	9.13		9.87		0.79	
Degrees of freedom	3		3		2	

*Coefficients significantly different from zero at the 5 percent level.

the case of column A in Table 5.10) minus the number of instrumented endogenous variables (one in the case of Column A in Table 5.10). The degrees of freedom for each overidentification test is shown below the statistic in tables 5.10–5.12.

For the first two instrumental variables regressions (for %GRID in column A and POPDEN in column B), the overidentification test rejects the assumption of orthogonal instruments at the five percent level. (The five percent critical value for chi-squared with three degrees of freedom is 7.81.) In column C, the overidentification test for the instrumental variables regression for RETDEN and SERVDEN (retail and service employment densities) does not reject the hypothesis that the instruments are valid. (The 5 percent critical value for χ^2 with two degrees of freedom is 5.99.) At this point, these results suggest that the instrumental variables technique is sometimes, but not always, an appropriate way to control for residential location choice when studying the link between land use and trip generation.

Step 2: San Diego (Census Tract Land-Use Variables)

The instrumental variables results for San Diego are shown in table 5.11.[23] The most important changes, comparing instrumental variables to the ordered probit regressions in table 5.5, involve the street grid variables (GRID and MIX) and the variable for commercial land use within each resident's census tract (%COMM).

Beginning with the results for the street GRID variables, GRID and MIXED, note that the coefficient on GRID is significantly negative in the instrumental variables regression in column A of Table 5.11 and that MIXED is insignificant in the same regression. Recall that both GRID and MIXED were significantly positive in the ordered probit routines in table 5.5. In terms of hypotheses, the ordered probit regressions indicate that grid-oriented neighborhoods generate *more,* rather than less, nonwork automobile travel. By contrast, the instrumental variables routine suggests that grid-oriented neighborhoods generate *less* nonwork automobile trips. This demonstrates how choices about econometric specification can crucially affect the interpretation of how land-use characteristics influence travel behavior.

The %COMM variable is significantly negative in column E of Table 5.11. Recall that, in the ordered probits in table 5.5, the only land-use mix variable that was significantly different from zero was the negative coefficient on %VACANT in column C. Thus, the only evidence on land-use mix from table 5.5 is that persons living in tracts with more vacant land take more nonwork car trips. In contrast, in the instrumental variables routines in table 5.11, the results suggest that persons living in tracts with more commercial land make fewer nonwork car trips. Again, the travel hypotheses of the new urban designs are only supported by the instrumental variables regressions.

Table 5.11: Instrumental Variables Regressions for Nonwork Automobile Trip Frequency (San Diego data)

	Column A: GRID & MIXED		Column B: HEAVY		Column C: Distance		Column D: %RESID		Column E: %COMM		Column F: %VACANT	
Independent Variables	Coeff	*T* stat	Coeff	*T* stat	Coeff	*T* stat	Coeff	*T* stat	Coeff	*T* stat	Coeff	*T* stat
Land use variables												
GRID (=1 for grid street network)	−1.3618*	−2.23										
MIXED (=1 for mixed street network)	0.7147	0.88										
HEAVY (=1 for dense street network)			−0.6595	−1.49								
D_CBD (distance from cbd)					0.1830*	3.22						
D_CBD2 (distance from cbd, squared)					−0.0054*	−3.06						
%RESID (percent residential)							−0.5091	−1.08				
%COMM (percent commercial)									−6.84857*	−2.51		
%VACANT (percent vacant)											0.6367	1.35
Income variables												
Household income <$20,000	−0.2867	−1.45	−0.2964	−1.54	−0.2705	−1.39	−0.3389	−1.80	−0.19789	−0.97	−0.3234	−1.71
Income between $20,000 and $40,000	0.1309	0.89	0.1761	1.26	0.2013	1.41	0.1764	1.27	0.22871	1.57	0.1776	1.27

(continued)

Table 5.11: (*Continued*)

Independent Variables	Column A: GRID & MIXED		Column B: HEAVY		Column C: Distance		Column D: %RESID		Column E: %COMM		Column F: %VACANT	
	Coeff	*T* stat	Coeff	*T* stat	Coeff	*T* stat	Coeff	*T* stat	Coeff	*T* stat	Coeff	*T* stat
Sociodemographic variables and controls												
Mean age of household members	0.0194*	3.85	0.0181*	3.97	0.2388*	4.56	0.1752*	3.86	0.01543*	3.25	0.0175*	3.85
Number of children < 16 years	0.7162*	6.23	0.6739*	6.23	0.7355*	6.43	0.6643*	6.13	0.65067*	5.82	0.6597*	6.07
Number of persons in household	0.6639*	7.44	0.7169*	8.68	0.6835*	7.98	0.7189*	8.71	0.67328*	7.74	0.7179*	8.68
Tenure (=1 if owner occupied)	0.2446	1.23	0.4266*	2.65	0.3892*	2.36	0.4616*	2.93	0.30633	1.74	0.4344*	2.70
SFR (=1 if detached single family)	−0.2167	−1.39	−0.1799	−1.20	−0.2250	−1.47	−0.1953	−1.31	−0.3468*	−2.11	−0.2177	−1.45
Constant	−0.4198	−0.99	−0.1907	−0.46	−1.6835*	−3.84	−0.3629	−0.95	0.21252	0.47	−0.7325*	−2.43
N	1320		1320		1320		1320		1320		1320	
R^2	0.26		0.31		0.28		0.31		0.27		0.31	
Overidentification test	6.34		10.30		1.06		11.35		5.41		10.69	
Degrees of freedom	2		3		2		3		3		3	

*Coefficients significantly different from zero at the 5 percent level.
CBD = Central business district; SFR = single family residence.

Theoretically, the instrumental variables technique should be preferred (contingent on the diagnostic tests discussed below), because that specification controls for the influence of land use on residential location choice. The implication is that, at least in the case of the San Diego data, the instrumental variables specification is vital in illuminating influences from land use that otherwise would have been missed in a regression analysis.

The results for commercial land use (column E of Table 5.11) are especially interesting. Recall from tables 5.8 and 5.9 that, for the San Diego data, commercial concentrations (as measured by %COMM) are negatively related to both median trip speed and median distance. These two effects work at cross purposes—slower speeds should discourage trip making while shorter distances might increase trip frequencies. Once residential location is controlled, the net effect in San Diego is to reduce nonwork car trip frequencies, as some proponents of mixed-use development have contended. Yet that result does not follow from a priori theory, and there could be different results in other urban areas. Similar tests should be implemented for other urban areas to examine how commercial land use influences trip frequencies in other locations.

The overidentification test statistics are again reported at the bottom of each column. Note that the overidentification test rejects the hypothesis of valid instruments in four of the six regressions.[24] This suggests some caution is warranted in using and interpreting the instrumental variables results, as many of the overidentification tests do not support the hypothesis of valid instruments.[25] Yet, importantly, the overidentification test does not reject the hypothesis of valid instruments for the regression for %COMM in column E of table 5.11, indicating that the intriguing results for that variable should be taken seriously.

Step 3: Orange County/Los Angeles (Zip Code Land-Use Variables)

The results from using the instrumental variables routine with land-use variables measured for ZIP codes for the Orange County/Los Angeles data are shown in table 5.12. Because of the labor-intensive nature of constructing the street geometry variable, %GRID was not measured for ZIP codes. Thus, only three of the four land-use variables for Orange County/Los Angeles were measured for ZIP codes.

The instrumental variables regression for population density (POPDEN) is reported in column A of table 5.12. The coefficient on POPDEN is insignificant, giving the same result as the ordered probit regressions with ZIP code land-use data. (Recall that the results of the ordered probits for ZIP code data, because they did not meaningfully differ from the ordered probits for census block group and tract land-use data, were not reported.) The overidentification test, reported at the bottom of column A, does not reject (at the 5 percent level) the hypothesis that the instruments are orthogonal to the error term.

Table 5.12: Instrumental Variables Regressions for Nonwork Automobile Trip Frequency (Orange County/Los Angeles data) ZIP Code Land Use Variables

	Column A POPDEN		Column B RETDEN & SERVDEN	
Independent Variables	Coefficient	T statistic	Coefficient	T statistic
Land use variables				
POPDEN (1,000 persons/sq mile)	0.0312	0.31		
RETDEN (1,000 retail jobs/sq mile)			−3.8203	−1.85
SERVDEN (1,000 service jobs/sq mile)			2.0006*	2.28
Income variables				
Household income ($1,000s)	0.0289	1.70	0.0456*	2.19
Household income squared	−0.0001	−1.35	−0.0002	−1.78
Sociodemographic and controls				
Female	0.7033*	2.53	0.4467	1.30
Age	−0.0260*	−2.09	−0.0375*	−2.42
Non-White	−0.0260	−0.06	0.1215	0.27
No high school	−1.8284	−1.16	−1.8440	−1.01
College	0.2317	0.80	−0.0340	−0.10
Number of children < 16 years	0.4854*	2.73	0.3773	1.75
Cars per drivers in household	0.1679	0.52	0.0072	0.02
Number of workers in household	0.2503	1.45	0.2331	1.15
Long commute	−0.0229	−0.04	0.2424	0.35
Work day	−0.9585*	−1.97	−0.6496	−1.06
(Long commute) * (Work day)	−0.3464	−0.56	−0.6626	−0.87
Constant	2.6309*	2.06	2.8348	1.78
N	434		432	
R2	0.11		0.11	
Overidentification test	7.38		0.09	
Degrees of freedom	3		2	

*Coefficients significantly different from zero at the 5 percent level.

The instrumental variables regression for RETDEN and SERVDEN, retail and service employment densities, is in column B. Note that RETDEN is significantly negative at the ten percent level and SERVDEN is significantly positive at the 5 percent level. Both RETDEN and SERVDEN, when measured for ZIP codes, were insignificant in the ordered probit regressions reported in table 5.4. The overidentification test, reported at the bottom of column B, does not reject the hypothesis of valid instruments.

Overall, the regressions in tables 5.10–5.12 suggest that, in certain instances, treating residential location choice (and thus land-use patterns near residences) as endogenous creates important changes in the results. In the case of the San Diego data and the ZIP code data for Orange County/Los Angeles, the instrumental variables results are more supportive of the travel hypotheses of the new urban designs than were the ordered probit regressions in tables 5.4–5.9.

The overidentification test provides reason for both optimism and caution. The overidentification test rejects the hypothesis of valid instruments in many of the specifications in tables 5.10–5.12. Yet the overidentification test accepts the hypothesis of valid instruments in some of the most intriguing regressions, and the results of the diagnostic test suggest that, in some instances, the instrumental variables approach is an appropriate way to handle location choice problems that are endemic in this research.

Most important, the results emphasize that the influence of land use is sensitive to choices about regression models and geographic scale, and researchers should consider that when examining the link between land use, urban design, and travel behavior.

Summary and Discussion

Both the theory in chapter 4 and the empirical examples in this chapter suggest that the link between urban design and travel behavior is a complex one. It is premature to conclude that, at the margin, neighborhood design can be a consistently effective transportation policy tool. But it is also premature to dismiss the possibility that land use does influence travel behavior. In fact, our regressions provide some evidence that street patterns and commercial concentrations are associated with fewer nonwork automobile trips. Yet those results became evident only when residential location choice and geographic scale were included in the statistical analysis.

More generally, our evidence is preliminary. Rather than firm policy prescriptions, these results attempt to disentangle the link between land use, urban design, and travel behavior. In that spirit, the most important lessons from the empirical work in this chapter are summarized below.

First, land-use and urban design proposals, if they influence travel behavior, do so by changing the price of travel. That idea should be the focus of future research on this topic. Linking neighborhood design characteristics to price variables provides a systematic framework that can guide empirical research and help structure its interpretation.

The importance of a price framework is illustrated by the regression results for commercial land use in San Diego in table 5.11. Individuals living in San Diego census tracts with larger proportions of commercial land use both have slower nonwork car trip speeds and take shorter nonwork automobile trips. Both effects are intuitive, both are predicted by many advocates of using land use as transportation policy, but (importantly) the net effect of both slower speeds and shorter distances on trip generation is ambiguous. Shortening trip distances can induce increases in trip generation, as discussed in chapter 4, while slowing travel speeds tends to reduce trip generation.

The regression results in Table 5.11 suggest that, in San Diego, persons living in tracts with more commercial land use make fewer nonwork car trips. That result depends crucially on the countervailing influences of slower trip speeds and shorter trip distances, and it would be naive to assume that the results in San Diego would hold for other urban areas. Instead, empirical research and policy practice in other urban areas should ask, first and foremost, how urban design influences average (or median) trip speeds and distances, and from there attempt to infer the net effect on travel behavior, traffic flows, congestion, and other transportation policy variables.

Second, geographic scale is important. The evidence for Orange County/Los Angeles revealed a link between land use and trip generation only when land use was measured for ZIP code areas. Urban designs emphasizing a "village" scale focus on small distances—typically a quarter mile or less. While there is evidence that such small distances are the appropriate scale for walking trips (Untermann, 1984), it is not clear, on an a priori basis, whether automobile trips are influenced by the urban form within small nearby neighborhoods or over larger areas. The evidence for Orange County/Los Angeles demonstrates the importance of examining different scales of geography when studying the link between urban design and travel.

Third, residential location choices matter. The evidence in this chapter supported the travel hypotheses of the new urban designs only when residential location choice was accounted for by the model. The point is not that incorporating residential location choice will reveal a link between urban design and travel in other urban areas—that is a topic for future study. Rather, the results of empirical research are sensitive to modeling choices regarding residential location. The instrumental variables technique outlined above is one way to control for the influence of residential location choice in trip generation. Future research should examine that approach further and should also adopt more detailed models of the joint decision about where to live and where to travel (as in, e.g., Linneman and Graves, 1983; Zax, 1991, 1994; Zax and Kain, 1991; Crane, 1996c; Van Ommeren, Rietveld, and Nijkamp, 1997).

Given that residential location choices seem bound up with individual preferences regarding travel behavior, an analysis of the travel impacts of urban design must consider how individuals choose where to live. This includes examining both the demand for different neighborhood types (location choice) and the supply of neighborhood types. The latter issue is important, but often overlooked.

Some authors suggest that modern planning regulations discourage the production of mixed-use, walking-oriented developments of the sort advocated by proponents of the new urban designs. The claim, best articulated by Levine (1998), is that zoning regulations restrict the supply of alternatives to the typical post–World War II, single-family, residential neighborhood. If true, the transportation claims of proponents of the new urban designs cannot be separated from the question of how to build neighborhoods that incorporate those designs.

The issue is not just whether persons travel differently once they live in neighborhoods designed differently, but also whether it is profitable and possible to build such neighborhoods. As one example, our research on transit-oriented development (TOD) in the following chapters suggests that, regardless of any effects on rail transit, local governments have incentives to avoid at least the residential component of TOD. This is but one take on the larger question of how private and public sector incentives influence the supply of neighborhoods that incorporate the new urban designs. Those issues are the focus of the next part of this book.

PART III
The Supply of Place

Chapter 5 closed by concluding that the supply of neighborhoods can be important in interpreting tests of travel/design linkages as well as the policy implications of those tests. A critical question is whether there are barriers to building communities that incorporate the new urban designs. If so, lowering those barriers could influence travel behavior even if the only effect is to provide a greater variety of neighborhoods, including those less dependent on car travel. The question of the supply of such neighborhoods is the focus of this and the next two chapters.

6 Neighborhood Supply Issues

Chapter 3 reviewed the literature regarding the influence of the built environment on travel behavior, and chapter 4 then described one way the issue might be usefully studied. The empirical work in chapter 5 provided intriguing results while illuminating some complex issues that remain unresolved in the analysis of urban design and travel behavior. Overall, our analysis thus far suggests that the link between the built environment and travel is intimately tied to the how urban form influences the *cost* of travel, and that the effect of design is *complex* in ways not adequately appreciated in most policy discussions. Neighborhood design in particular might affect automobile travel, but we still have much to learn about the nature, generality, and policy role of any such link.

That said, our analysis so far has been conventional in that it has focused on travel behavior. Yet that is only half of the story. It is also important to understand whether and how alternative land-use strategies might be more broadly implemented.

Having sketched out the role of the *demand* for travel in understanding the impacts of urban form on trip making, we now examine the *supply* of urban form. Put another way, how do communities shape cities toward transportation ends?

Neighborhood Supply

As discussed in chapter 3, a major difficulty in empirical work on travel behavior and urban design is that persons might choose residential locations based in part on how they wish to travel. Those who prefer walking are more likely to choose to live in pedestrian-friendly neighborhoods. People who prefer to commute by rail are more likely to live in transit-oriented developments.[1] If so, then simply looking at differences in travel patterns across different neighborhoods does not give insight into how urban design *causes* persons to travel differently. It is possible that urban design might not lead persons to travel differently at all, at least not in the sense of changing the way they desire to travel.

If there are an adequate number of communities providing less auto-dependent environments, then building more *might* have no influence on travel behavior. An "extra" transit-oriented neighborhood, in this sense, might attract residents who prefer to and then actually do travel by car for trips they could take by train. Travel behavior would then be largely unchanged by building more such places. The error of assuming this possibility away, discussed in chapter 5, is known as "self-selection bias." The observed differences in behaviors in different kinds of neighborhoods are explained by the self-selection of residents to those neighborhoods, not by the features of the neighborhoods themselves.

There is another possibility, however. Say the assumption that such communities are in surplus is plain wrong. Even if urban design does not influence individual preferences about travel, but transit- or pedestrian-oriented neighborhoods are constrained by local land use and design regulations, then building more developments that support alternatives to automobile travel can influence travel patterns simply by providing more places where people who want to drive less can do so.

The results in chapter 5 indicate that this may be an issue. Our evidence supporting a link between nonwork automobile trip generation and urban design came from regression specifications that corrected for the influence of residential location choice. Yet, rather than attempt to infer information about long-run residential location choices from regressions that explain short-run travel choices, we prefer to tackle the question of the supply of the new urban designs more directly.

In doing so, two issues are important. First, the travel pattern impacts of urban design flow both from (a) the link between those designs and long-run location decisions and (b) any shorter run link between urban design and how persons choose to travel. Second, to address the question of whether the new urban designs are undersupplied, the term "undersupply" must be defined in a more precise and policy-relevant way. We discuss each of those two issues below.

Any travel impact of the new urban designs is the result of effects in two markets. Building less auto-dependent neighborhoods can affect the residential location choices available.[2] Travel patterns could change either because people who desire certain types of travel prefer less car-dependent neighborhoods for that reason (an effect due to residential location), because urban design influences the way persons wish to travel (an effect that can be quite distinct from choices about residential location), or from some combination of both. Conceptually, the net travel impacts of urban design and land use result from effects that appear both in long-run residential (and other) location markets and in the shorter run market for day-to-day travel. The best way to analyze that behavior would be to model both the long-run location choice decision of persons and firms and short-run individual travel behavior decisions.

The question of the supply of the less car-dependent designs links the supply of particular neighborhood types to travel demand. Importantly, the link discussed here is between supply and demand in different markets. How does the supply of choices for long-run residential locations affect the demand for short-run day-to-day nonwork automobile travel?

More to the point, what does it mean to conjecture that certain urban designs might be undersupplied? At first glance, many would be tempted to answer that designs are in short supply if persons who wish to live in those neighborhoods cannot buy into them. In that sense, a supply shortage would have the intuitive connotation that there are not enough to accommodate all those who would wish to live there. Yet this definition of undersupply is not useful from either a policy or an analytical perspective.

Residential location choices, like many other things in life, are commodities bought at a price. Shortages are reflected in high prices, but a high price, in and of itself, need not be a policy problem. Luxury automobiles might be in short supply, in the sense that everyone who would like to have one cannot afford one. The same holds for beach-front homes selling for millions of dollars. If homes in less auto-dependent neighborhoods have a similar quality—if they are more expensive than some persons might like—why would that be a policy problem? We argue that, as long as the market for residential development operates efficiently, the supply of the new urban designs is not a policy problem, and any scarcity of that good should be no more troubling than scarcities of luxury cars or seven-bedroom houses.[3]

The question of the supply of the new urban designs then becomes a question not of the level of supply, as such, but of how closely the housing market approximates the characteristics of a well-functioning competitive market in relation to transportation-oriented communities. Are there factors that constrain the supply of these designs, *and* are those factors due to imperfections in either the functioning or the regulation of the housing market? That is the policy question that must be addressed, since any undersupply that results from market failure or from inappropriate regulation is appropriately a policy issue.[4]

Market Failure and Government Failure

Different persons often desire to live in different types of neighborhoods, so supply in the market for neighborhood types is inherently a matching problem—how well does the array of available neighborhood types match the diversity of neighborhood types demanded? With that question in mind, consider constraints on the supply of neighborhood types from two sources: *failures in the market for residential development* and *failures of government regulation.*

In terms of market failure, the question is whether the housing market, acting on its own, will provide the diversity of neighborhood types that correctly matches the variety in consumer demand.[5] In other words, would an unregulated housing market respond to the wishes of consumers who seek to purchase homes in places that adhere to transit- and pedestrian-oriented guidelines? We see no clearly articulated reason why an unregulated housing market would not. There is some concern about whether developers would build communities in ways that depart from more traditional practice, or whether banks would lend money for such developments (Fulton, 1996). Yet whether developers and banks will or will not, in time, build alternative designs to meet market demand remains to be seen. Certainly many developers say they plan to, and many projects are in place, but how well they incorporate alternative transportation plans is not generally known.[6]

If the development market is not a source of economically inefficient supply constraints, what of governments? Is there government failure of one sort or another, rather than market failure (e.g., as in Wolf, 1993)? Some conclude so.

Levine (1998), for example, argues that zoning regulations are often at odds with the less car-dependent places.[7] This idea is echoed more stridently in the writing of Kunstler (1993) and was also developed in the widely cited, and highly regarded, work of Jacobs (1961). At face value, there is much in local zoning codes that appears to work that way. Many zoning codes are grounded in attempts to segregate residential from commercial, office, and industrial land uses, rather than mix them. The New Urbanism often proposes densities that are higher than common in most suburban areas, and many zoning codes have maximum densities that cannot be exceeded. The street design standards of the Neotraditional designs of Duany and Plater-Zyberk (1991, 1992), for example, emphasize narrow lanes, pedestrian access, and possibly design elements that are intended to slow traffic. These can be at odds with local street codes that focus on facilitating automobile traffic flow.

The disjunction between the ideas of the new designs and existing traffic codes is serious enough that the Institute of Transportation Engineers (1997) recently developed guidelines to help cities modify their traffic codes to allow the narrower, more pedestrian-friendly streets that are part of these plans. Overall, it is not difficult to see how zoning (and traffic) codes might restrict rather than facilitate the development of some neighborhood designs.

The question is then not only *whether* local regulations constrain the supply of the new urban designs, but *why* they do so. There are, broadly speaking, two types of answers, one more benign than the other. First, zoning and traffic codes reflect earlier thinking that segregated land uses and deemed that automobile mobility should be essentially the entire focus of local land-use regulation (see Altshuler, 1965). To that extent, local regulations might simply have failed to keep pace with changing individual preferences. Education about how development options have changed, and how local land-use regulation should evolve in ways that do not constrain those options, would presumably go a long way toward overcoming the problem of local governments unwittingly restricting less car-dependent communities.

The point is not that such communities should be built but rather that they should not be unnecessarily denied permission to build. Certainly, that education should occur, and we believe it is occurring in many urban areas. Many suburban communities are experimenting with mixed-use developments, traffic calming, and improved pedestrian access (Knack, 1995, 1998; Bernick and Cervero, 1997). This is different from saying that such communities can deliver on their transportation promises by rhetoric alone. The lessons of chapters 3 and 5—that is, careful case-by-case attention to traffic impacts, and the preparation of contingency plans if traffic does not respond as expected—still apply. But if all that is needed is more mention of the range of neighborhood possibilities that are available, then we are on our way there.

The other possibility, which leads to less sanguine conclusions, is that local governments have incentives to regulate in ways that constrain this type of project. If so, this is more than an education problem. There are many possible reasons why local governments would behave this way. Many of these have been discussed in other contexts, and a brief review is appropriate before proceeding (Fischel, 1985).

For one, it has long been contended that local governments might engage in *exclusionary zoning*—attempting to keep out persons of certain races or socioeconomic classes by requiring, for example, minimum lot sizes that essentially price many of the targeted persons out of the city (e.g., Danielson, 1976). The practices of *fiscal zoning,* such as minimum lot sizes, are often the same as the techniques of exclusionary zoning, but the motive for fiscal zoning is more directly related to local taxes and expenditures (e.g., Mills and Oates, 1975). In a system of property tax finance, municipalities might seek to discourage in-migrants who either consume large amounts of public services or desire low-cost housing that would lower the per-capita property tax base. In practice, it can be hard to distinguish exclusionary zoning from fiscal zoning because in both instances rich communities will try to use zoning to discourage poorer in-migrants from moving to the wealthy city. Yet the motives for exclusionary and fiscal zoning differ, and fiscal zoning is best considered one among several possible motives for exclusionary zoning.

Both exclusionary and fiscal zoning are strategies by which communities pursue their narrow self-interest to the possible detriment of broader social goals. Such social goals might well include the well-being of potential in-migrants who prefer to live in the exclusionary city but who are kept out by regulatory policy. If the gains to the potential in-migrants exceed the losses that accrue to city residents when, for example, the per-capita property tax base is lowered by in-migration, the city's attempts at fiscal zoning can be undesirable from a social perspective even while those actions further the narrower self-interest of the city residents.[8]

For our purposes, does a similar phenomenon occur in relation to alternative transport neighborhood designs? Might local incentives encourage municipalities to constrain the supply of certain neighborhood types in ways that work against broader social goals? If so, then local incentives embedded in land-use regulation must be understood both to assess whether the supply of some neighborhood styles is artificially constrained and to point the way toward any needed policy reforms.

There are several reasons why localities might use their zoning power to inhibit specific kinds of neighborhood developments. For example, local governments might zone for fiscal or economic reasons—to increase their tax base or to lure new jobs—and those zoning policies might not be consistent with less auto-oriented developments. In some communities, increasing density can be politically sensitive, which by itself could provide an impediment to the new urban designs. The street traffic implicit in the new urban designs might create concerns about safety in suburban communities that are not accustomed to pedestrian environments (see the discussion in Handy, Clifton, and Fisher, 1998).

All of these are reasons why local regulations might constrain urban form and land use in ways that are suboptimal in the very specific sense that developments which would otherwise be built by private markets are not. We do not examine all these issues here, but as an illustration, we do consider a specific set of issues regarding a specific type of transportation-oriented land-use strategy: Do fiscal and economic concerns restrict the supply of transit-oriented development (TOD)? We use that topic both as a window into broader issues and as a way to illustrate how empirical techniques on this topic should focus on illuminating the behavior of local governments.

Planning Incentives

Do municipalities, acting in their own self-interest, have incentives to restrict the diversity of neighborhood types? Our answer is possibly yes, and we examine this issue in some detail in chapters 7 and 8. Yet the question of how local governments might constrain urban design is quite broad.[9] Rather than looking at all possible local incentives and all types of designs, we ask whether local governments might restrict the supply of transit-based housing for fiscal or economic reasons. We believe this single example can nonetheless clarify a broad range of issues concerning the supply of the new urban designs.

Transit-based housing is a key component of TOD, an idea first popularized by Kelbaugh (1989) and Calthorpe (1993). TODs are pedestrian-friendly, mixed-use developments focused around rail transit stations. They are typically built at higher densities than most suburban development and emphasize public spaces and aesthetically pleasing streetscapes that encourage foot traffic. While TOD and the New Urbanism developed separately, they share many characteristics in that TOD is, in many ways, New Urbanist neighborhoods built around rail transit stations.

Examining transit areas, rather than other new kinds of neighborhood designs, has several advantages for our purposes. First, the potential sites for those neighborhoods can be easily identified, as they focus on rail transit stations. Second, many urban areas in the United States have been building or planning rail transit systems, so many have several potential sites that can be the basis for empirical analysis. Third, local fiscal and economic issues are especially stark in the case of TOD.

In some cities, such as Washington, D.C., rail stations have become centers of economic development (Cervero, 1994a, 1994b). For that reason, rail transit is possibly a venue over which battles for local economic development are fought. Transportation investments have often fueled debates about their effect on municipalities that compete for mobile economic activity.[10] Thus, the economic pressures leading municipalities to compete for jobs might be especially stark around rail transit stations.

Development near rail stations can also have fiscal impacts. In a system of local property taxes, residential development might not pay its own way, in the sense of bringing increased tax revenues that equal or exceed the increased local expenditures (schools, public safety, and the like) required to serve the new residents (e.g., Ladd, 1975; Ladd and Yinger, 1989). Commercial development, on the other hand, often does not bring the same public service requirements, and so might be more attractive to local governments concerned about the fiscal impacts of growth. With local sales tax finance, these trends are exacerbated, due to the fact that commercial properties generate taxable transactions.[11] With either local property or sales tax finance, commercial development might be favored over residential development for fiscal reasons.

In our own work, we have suggested that those fiscal pressures lead municipalities to favor commercial (or office) development near rail stations, to the detriment of the residential component of TOD that is a vital part of the mixed-use character of those neighborhoods (Boarnet and Crane, 1997, 1998a; Boarnet and Compin, 1999). The argument, loosely speaking, is that if local governments prefer commercial or office development for fiscal reasons, they might especially prefer that type of development near rail stations, given the fact that rail stations can be a natural place to focus new commercial and office development. We repeat that analysis in chapters 7 and 8 by examining both the economic and the fiscal incentives for commercial development near rail stations. If local governments are motivated in part by economic and fiscal concerns, and if those motives influence local land-use regulations, then the effect of those motives should be evident in land-use regulation near rail transit stations.

Again, this is a window into a broader issue. Do local governments have incentives to zone in ways that restrict the supply of some urban and suburban designs? Examining that question would be difficult, in part because of the difficulty in operationalizing and measuring the various characteristics of alternative design schemes and in part because it is difficult to identify where any such restrictive zoning would be most evident. Instead, it is easier to measure zoning that might constrain residential TOD, and to identify where those constraints should be evident (near rail transit stations). Our point is not that if local zoning restricts residential development in transit station areas then it will necessarily also restrict other designs. Rather, there is a need to understand how municipalities regulate the land market and, more specifically, how parochial self-interest can provide incentives to restrict the supply of particular types of communities. By examining that issue within the context of TOD, we illustrate both empirical techniques and policy ideas that can have broader applicability to other types of community plans.

Policy and Neighborhood Supply

Some characterize the New Urbanism as an ambitious government intervention in land markets (e.g., O'Toole, 1996). If so, as in chapter 2, we would wonder what market failures justify it. But what if government regulations prohibit such developments, rather than require them? Levine (1998), in particular, has argued that the new urban designs should be viewed as *deregulation,* not regulation, and that accordingly advocates of free market policies should favor lessening of rules that facilitate the development of less car-dependent places.

First, the question of whether the new urban designs are constrained by government regulation is an empirical one, on which there is little evidence. But our analysis in chapters 7 and 8 provides at least some reason to believe that local incentives can at times work against the development of new urban designs, consistent with the evidence that Levine (1998) presents.

More important, focusing on the supply of neighborhood types makes some policy decisions easier. If someone wishes to live in New Urbanist or similar neighborhoods, there is value in having such neighborhoods exist. If, further, it can be demonstrated that such designs are undersupplied due to local government regulations, there can be benefits in relaxing those regulations.[12]

What would happen if regulations were relaxed? In many instances, this question need not be answered a priori. If the development market otherwise functions well but is constrained by onerous or inappropriate local regulations, relaxing those regulations permits the market to respond to latent demand—be the market niche large or small.

Having said that, it is useful to have some sense of the likely effect that relaxing local regulations will have on development patterns, urban areas, and automobile travel. Can the solution to the policy problem described in chapter 2 be as simple as removing government constraints to certain subdivision features?

We suspect not, for two reasons. First, while there may be latent demand for such developments, there might not be enough to result in the large-scale changes in urban form needed to have a sizable impact on metropolitan travel patterns.[13] Second, the travel impacts of building more neighborhoods that are less car dependent are in important respects still incompletely understood. For both reasons, policy analysts should view facilitating the development of the new urban designs as one part of a broader menu of possible transportation policies.

The policy implication of focusing on the supply of neighborhood types is thus twofold. If local government policies and incentives constrain the development of the new urban designs, planners should recognize that facilitating certain development patterns might require deregulation rather than additional regulation. This changes some of the focus of policy attention. Second, if as chapters 7 and 8 indicate, municipalities have incentives to resist some land uses aimed at reducing car travel, the implementation problems associated with building those developments can be considerable and were underestimated in the past.

Proponents of high-density residential development near rail stations argue that such projects will get more people onto trains, reduce developers' expenses, and lower commuting costs, housing prices, and air pollution in the bargain. This chapter focuses on a separate question: Transit-oriented housing developments remain relatively rare. Why?

Our analysis of more than 200 existing and proposed southern California rail transit stations concludes that cities' parochial fiscal and economic interests may be an obstacle—and an opportunity.

7 Transit-Oriented Planning

There has been a boom in American rail transit construction in the past two decades. That new investment has prompted the question of what planners can do to support rail transit. One popular answer has been transit-oriented development (TOD),[1] increasingly described as a comprehensive strategy for rail-based land-use planning throughout an urban area. This is most clearly illustrated by Bernick and Cervero's (1997) description of how such projects can link together to create "transit metropolises" where rail is a viable transportation option for many of the region's residents.

In addition, TOD provides an opportunity to examine the regulatory issues discussed in chapter 6, both because it is an explicit attempt to use urban design as transportation policy and because the intergovernmental issues are especially stark in relation to these developments. Having discussed how travelers behave in the first part of this book, we now ask what we know about how cities behave.

Stated in general form, the question is rather broad. It concerns the process by which cities and other land-use authorities decide where to put streets, how to structure the local hierarchy of streets, when to develop more or less densely, how to position employment centers relative to residential areas, and so on. Still, the feasibility of land-use plans with transportation goals depends critically on how such authorities behave. Any discussion of the effectiveness of these strategies must address both how communities plan for transportation and how travelers respond to those plans.

Transit-Oriented Planning

The primary transportation goal of TOD generally, as currently practiced, is to coordinate land-use policies to support rail transit. In particular, focusing both residential and commercial development near rail transit stations is aimed at increasing rail ridership (e.g., Bernick, 1990; Bernick and Hall, 1990; Calthorpe, 1993; Cervero, 1993; Bernick and Cervero,

1997). Some evidence suggests that residents near rail transit stations are two to five times more likely to commute by rail when compared with persons living elsewhere in the same urban area (Pushkarev and Zupan, 1977; Bernick and Carroll, 1991; Cervero, 1994d).[2]

Significantly, the emergence of transit-oriented plans can be traced in part to development experiences near San Francisco Bay Area Rapid Transit (BART) stations. When BART opened in the early 1970s, planners assumed that the new rail transit stations would become centers for economic development or (more often) redevelopment. The presence of the heavy rail transit system would encourage medium- and high-density development, and presumably some of that development would be residential units offering easy access to a BART station (Bernick, 1990). Yet early evaluations suggested that BART created at best only small land-use effects (Webber, 1979), and later studies still found only small effects (Cervero and Landis, 1997; Landis and Cervero, 1999).

By the late 1980s, it was clear that redevelopment near many BART stations had proceeded at a slower pace than expected. Planners began to conclude that the land market, if left to its own devices, would not fully exploit the development opportunities near stations. Some practitioners and scholars argued that government would have to intervene. Suggested policies included rezoning land near stations for residential uses, offering density bonuses or subsidies, or otherwise facilitating development (especially residential development) near rail transit. This policy activity meshed with academic thought that advocated a return to more dense pedestrian- and transit-oriented communities (e.g., Kelbaugh, 1989), and the TOD idea was born (Bernick, 1990; Bernick and Cervero, 1997; Culthorpe, 1993).

Yet for several years residential transit-oriented planning was somewhat rare in practice, so much so that some supporters regard the lack of residential development near rail transit stations as a public policy problem requiring both explanation and intervention (Bernick, 1990; Bernick and Hall, 1990, 1992; Calthorpe, 1993). The supply issues discussed in chapter 6 are thus potentially important for this design strategy.

Many TODs involve a multitude of public and private actors, yet the incentives of some are often overlooked. To be built, these developments require both a healthy local economic environment and supportive municipal and regional land-use regulation. In this chapter, we focus on an empirical analysis of the incentives of local governments, which have virtually all land use regulatory authority in most states in the United States. Because TOD requires that residential, commercial, and mixed-use projects be built near rail transit stations, local land-use regulations can be crucial in either facilitating or impeding those developments. This issue applies equally well to all urban designs, but particularly those requiring changes in land-use policy.

We suggest three reasons to doubt whether municipalities, if left to their own devices, would aggressively pursue transit-based housing. The first is financial, based on the increasingly tough economic competition between cities, within and across metropolitan areas. The second is historical. Advocates of transit-based housing often ignore the economic and political forces that led to the demise of notable earlier rail transit systems; yet those same economic and political forces are alive and well today and may cause localities to shy away from transit-based housing. The third reason is an example, recounted below, of what happened when several suburban city governments had the opportunity to provide the initial impetus for an urban rail plan. The choices those municipal governments made about station sites provide instructive lessons for advocates of transit-based housing.

Economics

One explanation for the limited implementation of transit-oriented housing is that localities aim, by way of either long-term planning strategies or incremental zoning decisions, to use rail transit stations as a means to enhance their fiscal position. To the extent that rail transit stations are perceived as opportunities to focus new development or redevelopment, a municipality might choose to emphasize land uses that have the most favorable impact on its tax base.

Planners have long recognized that urban and suburban municipalities tend to compete for both employment share and tax base. The term "fiscal zoning" has even been coined to describe zoning that is aimed at boosting a city's revenue, and the use of fiscal impact analysis to evaluate the merits of one land use over another is now widespread (Wheaton, 1959; Burchell and Listokin, 1980; Burchell et al., 1998).

Many authors have suggested that local governments will favor commercial over residential uses in a system depending on either property or sales tax finance, since commercial properties often generate tax revenue without the service requirements of new residents (e.g., Ladd, 1975; Schneider, 1989; Altshuler and Gómez-Ibáñez, 1993). In California, the propensity to use fiscal zoning was arguably exacerbated after the property tax limitation Proposition 13 passed in 1978. That act fixed the typical property tax rate at near 1% of assessed value, and set assessed property value at the higher of either (a) its 1978 market value plus a maximum of 2% appreciation per year, or (b) its last sales price since 1978 plus a maximum of 2% appreciation per year. (Since the early 1980s communities have also been permitted an additional special assessment—or "Mello-Roos"—of up to 1% of assessed value explicitly for the purpose of financing infrastructure associated with new developments.)

Proposition 13 decreased the relative importance of the property tax and increased the relative importance of the sales tax as local revenue sources (Lewis and Barbour, 1999). (A share of locally generated sales tax collections is rebated by the state back to each local jurisdiction.) Localities in California do not have much direct control over either the property or sales tax rate.[3] Their influence over the fiscal environment is, rather, mainly indirect via their control over the revenue-generating ability of alternative land uses.

The broader trends are by no means unique to California however. A general pattern of decreased federal funding for municipal operations over the past two decades, and the labor-intensive nature of many local services have led to growing state and municipal fiscal stress nationwide (Ladd and Yinger, 1989; Gold and Zelio, 1990). Municipal responses have ranged from service cutbacks, on the spending side, to increased use of debt (Crane and Green, 1989) and impact and application fees aimed at getting development to "pay its own way" (e.g., Nelson, 1988; Altshuler and Gómez-Ibáñez, 1993; Levine, 1994; Brueckner, 1997b; Dresch and Sheffrin, 1997; Burchell et al., 1998). Planners have long incorporated fiscal impact analysis as either a formal or informal element of the development evaluation process, and both common sense and anecdotal evidence suggest that, if anything, this will continue as fiscal pressures rise.

History

In the years between World War I and World War II, Los Angeles's renowned urban rail system lost patronage to the automobile and began to fall into disrepair. After World War II, Los Angeles was at a crossroads, having to decide what to do with its urban rail while embarking on an ambitious program to build freeways. As Adler (1991) has documented, the city's answer was to some extent preordained by the perceived advantages of auto travel. Rail transit was probably destined to lose its dominant role, and in the post–World War II years the question was what, if any, rail service would be preserved in the West's largest metropolis. The answer, it turned out, was none.

Despite the popularity of blaming coalitions of oil companies and automobile manufacturers for the demise of the Pacific Electric Railway (and likewise the other rail lines in Los Angeles), Adler (1991) shows that their disappearance is best credited to the workings of political coalitions that favored freeways over rail. For many suburban municipalities, the advantage of a highway network was that it supported economic development within their communities. Rail, with its hub-and-spoke orientation, was perceived to favor the economic development of downtown Los Angeles. Although concentrating business and commercial activity in the central core appealed to the downtown business community, it was anathema to the developing economic centers in places such as Santa Monica, the San Fernando Valley, the San Gabriel Valley, and Long Beach.

In the end, highways drew support from a broad coalition of suburban municipalities and downtown business interests (Wachs, 1984; Adler, 1991). Most major political actors viewed freeways as supporting economic development in their communities, while rail was perceived as supporting growth only in the downtown. It was as if local governments voted in their own economic development interests, and more municipalities perceived freeways as benefiting their local economies. The politics of local economic development helped shape a transportation system, and in the process led Los Angeles from rail to freeways. Given this history, and the fact that political battles over transportation often are influenced by the spatial pattern of economic benefits, it is reasonable to expect that the current generation of rail transit systems in Los Angeles and other regions is subject to the same political pressures.

A Suburban Example

Forty years after the decisions that led to the demise of Los Angeles's Pacific Electric rail system, the southern California region began to reconstruct the rail transit system that it no longer had. While most of the activity was focused in Los Angeles (and, even earlier, in San Diego), other areas within the metropolis also pursued rail transit plans. Orange County, a densely developed residential and employment center to the south of Los Angeles, began to seriously consider an urban rail system in the late 1980s.

The original Orange County plan was developed not by county government, but by a coalition of cities (Central Orange County Fixed Guideway Project, 1990). Those cities—Anaheim, Costa Mesa, Irvine, Orange, and Santa Ana—are of comparable size and political influence. They are also located along a north–south line containing the county's most dense development. The original project prospectus suggested using an elevated fixed guideway (e.g., a monorail), greatly reducing right-of-way acquisition costs and making right-of-way a less important issue in siting both the line and the stations.[4] Having been freed from some of the most difficult siting constraints, system planners were not tied to building rail transit on historical routes or on current freight lines. With the ability to site the line without concern for right-of-way issues, it is telling that the five Orange County cities chose to use their rail system to connect employment centers, rather than residential centers[5] (see the proposed line placement in figure 7.1).

The coalition cities perceived urban rail as a catalyst for local economic development. While some thought was given to how to link residential areas to the transit system, mainly by the road network, the early planning documents and proposed station locations indicate that this question played a secondary role in the station siting decisions:

> The changing character of land use patterns in Orange County over the past two decades has influenced the development of employment and retail commercial concentrations in growing activity centers. An activity center is defined as an area of existing and increasing intensification of office and commercial activity with a minimum of 13,000 employees and an employment density of 10,000 jobs per square mile. . . . As proposed, the [rail transit] Project will serve [the 13 existing activity centers] on the main line and the circulation systems. . . . Since the [rail] Project will not provide all the necessary links to residential areas to accommodate the home-to-work commute, the system will be integrated with the countywide HOV system of commuter lanes and transitways now being developed. . . . (Central Orange County Fixed Guideway Project 1990; pp. 3, 5)

PROPOSED ORANGE COUNTY URBAN RAIL LINE

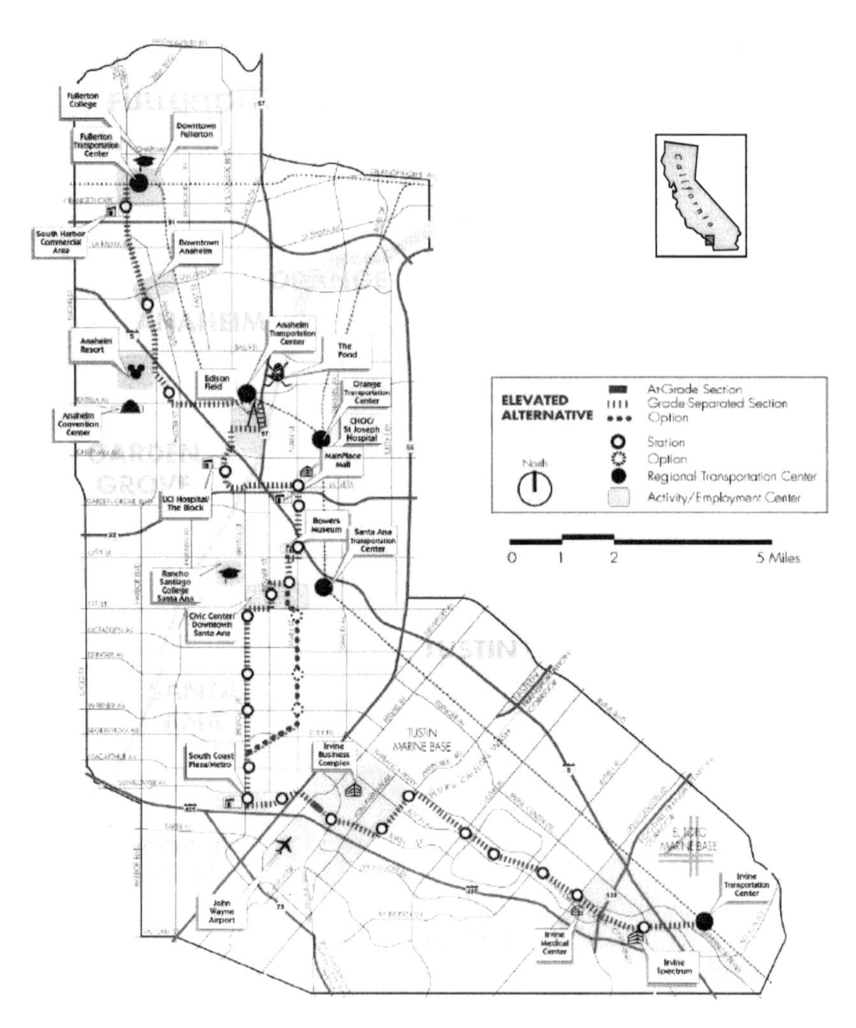

Figure 7.1. Proposed light rail route map of the Central Orange County Fixed Guideway Project (1999), linking major activity areas.

Rail would be a tool for local economic development, and stations were proposed for the county's largest commercial and employment centers.

Evidence from Southern California

The Orange County story fits the idea that localities will view development near rail stations as a way to enhance their economic or fiscal position. One implication is that local fiscal incentives toward transit station areas, heretofore often overlooked, should be examined more generally. It matters little whether neighborhood design influences travel behavior if those designs overlook, and thus underestimate, political and regulatory challenges in implementation.

The purpose of the empirical analysis below is not to comprehensively examine all possible political obstacles to TOD implementation, but instead to focus on the role of economic and fiscal incentives. Two questions are important: Do localities regulate land near rail station areas in ways that suggest systematic local incentives motivated by the economic and fiscal concerns described earlier? And if so, what is the implication for TOD and, more generally, for the role of urban design in transportation policy?[6]

There are four transportation authorities currently operating rail transit in southern California, and in 1994 we collected data on all 232 existing and proposed rail transit stations in the region at the time (see table 7.1 for a summary). The oldest existing system is operated by the Metropolitan Transit Development Board (MTDB) in San Diego, best known for its trolleys running from downtown to the Mexican border. The Los Angeles County Metropolitan Transportation Authority (MTA) opened the Blue Line in 1990, the Red Line (the city's first subway) in 1993, and the Green Line in 1995. The Southern California Regional Rail Authority (SCRRA) operates the MetroLink interurban system, which opened in the early 1990s. The North County Transit District (NCTD) operates the Coaster, which opened in 1997 and runs from northern San Diego County into San Diego.[7] As mentioned, the Orange County Transportation Authority (OCTA) is engaged in the planning process for a rail transit line.

We also gathered zoning data for quarter-mile-radius circles centered on each of the region's rail transit stations that were open or proposed at that time. We chose zoning as the best available measure of local land-

Table 7.1: Operating Status of Stations in Southern California

Operating Authority and Line Date	No. of Stations in Dataset	No. in Operation as of 1995	Proposed Start Date		
			Late 1995	Later	No Star
MTA	70	26	13	26	5
Blue	35	22	0	13	0
Red	18	4	0	13	1
Green	17	0	13	0	4
MTDB	41	35	6	0	0
East	15	12	3	0	0
North	3	0	3	0	0
Center	11	11	0	0	0
South	12	12	0	0	0
NCTD	22	0	6	16	0
OCTA	44	0	0	0	44
SCRRA	55	47	0	8	0
Total	232	108	25	50	49

use regulations. This required collecting zoning information from the eighty municipalities that have land-use authority over part or all of a station's quarter-mile area in southern California.

Zoning data were gathered for six categories: (1) low-density residential, (2) high-density residential, (3) all residential, (4) commercial, (5) mixed use, and (6) industrial. Close attention was required to categorize zoning data, which are not necessarily consistent across municipalities. For example, discrepancies in the ways cities classify and report low- and high-density residential cause the "all residential" category to be not the simple sum of "low-density residential" and "high-density residential." In addition, many municipalities either do not have a "mixed-use" zoning classification or do not report the percentage of municipal land that is zoned for mixed-use development. Thus, the "mixed use" data are missing for many stations and lines.

Having said that, a simple comparison by rail line does provide an interesting look at how the patterns of land-use zoning near stations are related to station characteristics. The zoning data are provided in two forms. Table 7.2 shows the share of land within a quarter mile of a station in each zoning category ("station share"), and table 7.3 shows the ratio of the station's land in any zoning category divided by the share of the surrounding city's land in that same zoning classification ("station ratio"). These measures are explained further below. Each captures a different aspect of the relationship between zoning near a station and the community containing that station.

Zoning Measure 1: Station Share

Looking at the "station share" measure for the zoning categories in table 7.2, and as illustrated by transit authority in figure 7.2, on average 9.5 percent of the station area land is high-density residential, 24.5 percent is residential, 22.5 percent is commercial, 6.0 percent is mixed use, and 14.9 percent is industrial. Residential is the highest single category, most of which appears to be low density.

This pattern differs substantially by rail line, however. By line, the "low-density" share varies from 1 percent for MTDB Center, in downtown San Diego, to 26 percent for the two MetroLink lines to Moorpark and Hemet—both peripheral suburban communities. On average, the NCTD stations are the most "low-density residential" and the MTA stations the least. These figures accurately reflect the range one expects when comparing suburban residential communities with major central employment centers. The pattern of "high-density" residential is quite different, with MTA stations having the highest share at 17 percent and OCTA and MetroLink stations the least at about 5 percent.

The share of station area land zoned commercial across rail lines ranges from 1 percent in MTDB North stations, to 46 percent for the main OCTA stations. Most lines are in the 15 percent to 30 percent range, however. The most commercial station areas are on the OCTA lines, on aver-

Table 7.2: Land Use Zoning Near Rail Stations, Station Share

Rail Line	Low-Density Residential	High-Density Residential	All Residential	Commercial	Mixed Use	Industrial
MetroLink total (SCCRA)	15.7%	5.1%	20.9%	20.0%	2.7%	25.4%
Hemet–Riverside	26.2%	4.6%	30.8%	16.5%	—	16.2%
Moorpark–L.A.	26.5%	10.4%	36.9%	12.4%	—	25.9%
Oceanside–L.A.	4.2%	6.9%	11.1%	19.6%	8.3%	25.6%
Redlands–San Bernardino	4.2%	0.0%	4.2%	41.8%	4.4%	3.3%
Riverside–L.A.	15.0%	1.0%	16.0%	19.3%	—	39.0%
San Bernardino–L.A.	19.3%	6.3%	25.6%	19.0%	2.2%	26.7%
Santa Clarita–L.A.	20.6%	3.2%	23.7%	26.1%	5.1%	16.7%
San Bernardino–Riverside	5.0%	14.7%	19.7%	16.2%	—	27.2%
MTA total	12.4%	16.8%	29.1%	21.1%	7.1%	11.6%
Blue	11.9%	17.3%	29.2%	14.9%	9.3%	14.2%
Green	18.3%	10.2%	28.5%	16.2%	9.4%	16.5%
Red	6.7%	21.3%	28.0%	38.1%	—	3.3%
MTDB total	14.5%	7.3%	21.9%	16.3%	7.0%	11.0%
Centre	1.0%	0.2%	1.2%	21.7%	8.7%	19.1%
East	22.5%	16.7%	39.2%	18.8%	4.0%	6.3%
North	5.3%	0.0%	5.3%	0.7%	13.2%	7.8%
South	19.3%	4.0%	23.3%	12.1%	7.7%	10.3%
NCTD	23.1%	9.1%	32.2%	19.1%	12.5%	13.4%
OCTA total	14.6%	5.4%	20.1%	35.1%	4.3%	11.3%
Main	10.2%	4.7%	14.9%	46.4%	2.0%	3.8%
Alternate	24.2%	6.8%	31.0%	18.3%	7.9%	11.9%
Extension	14.3%	5.6%	20.4%	28.5%	5.4%	24.7%

Table 7.3: Land-Use Zoning Near Rail Stations, as the Ratio of the Share of Land in Each Use Within 1/4 Mile of Station Divided by the Share of City Land in That Use: Station Ratio

Rail Line	Low-Density Residential	High-Density Residential	All Residential	Commercial	Industrial	Mixed Use
MetroLink total (SCCRA)	1.67	1.97	0.57	4.37	5.23	—
Hemet–Riverside	0.83	2.20	1.07	5.16	8.49	—
Moorpark–L.A.	0.86	4.52	0.98	9.21	4.06	—
Oceanside–L.A.	0.18	1.73	0.30	5.30	4.44	1.29
Redlands–San Bernardino	0.10	0.00	0.10	5.03	0.29	—
Riverside–L.A.	0.48	0.11	0.45	3.27	7.23	0
San Bernardino–L.A.	6.23	2.68	0.56	2.18	3.99	0
Santa Clarita–L.A.	0.55	0.00	0.52	4.73	2.34	—
San Bernardino–Riverside	0.07	13.61	0.96	3.82	5.64	—
MTA total	0.44	0.87	0.56	1.95	1.78	—
Blue	0.47	0.97	0.62	1.28	2.17	0
Green	0.63	0.53	0.44	1.86	0.89	0
Red	0.21	0.89	0.51	3.14	1.72	0
MTDB total	0.62	1.38	0.72	3.56	2.96	—
Centre	0.06	0.02	0.05	5.57	5.37	—
East	0.66	3.51	1.08	3.00	1.73	1.14
North	0.31	0.00	0.21	0.19	2.19	—
South	1.29	1.15	1.05	3.19	2.38	—
NCTD	0.68	4.29	0.83	4.72	4.26	10.99
OCTA total	0.31	0.88	0.40	8.37	1.35	11.85
Main	0.22	0.48	0.31	14.11	0.33	8.34
Alternate	0.45	0.72	0.56	1.32	1.33	17.80
Extension	0.35	1.39	0.43	3.71	3.22	15.25
All	0.76	1.47	0.59	4.41	2.93	8.61

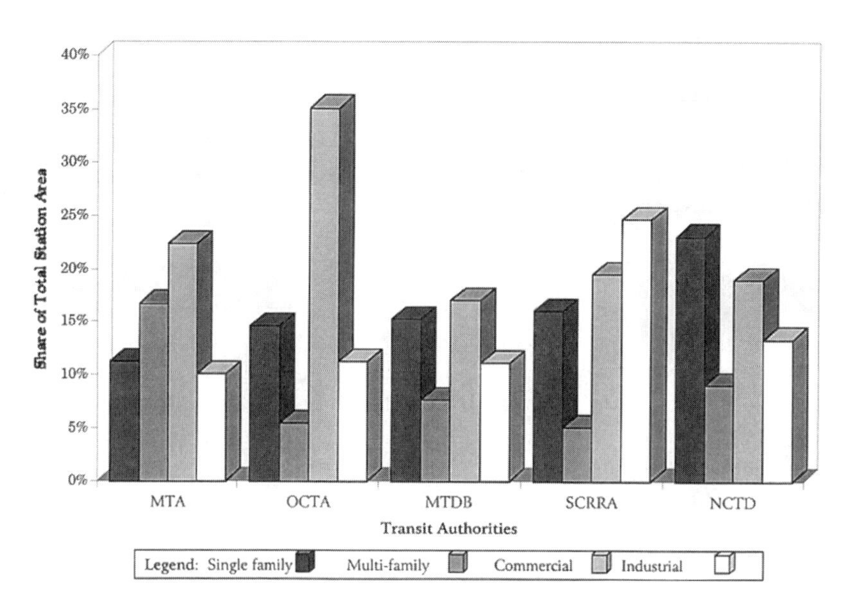

Figure 7.2. The share of station area land zoned in each category, by transit authority ("station share").

age, and the least commercial are on the MTDB stations. Though less station area land is commercial than residential, on average, the difference is not great. Combining commercial land with the land zoned for industrial use, more land in the station area would appear generally to be for shopping or employment generation purposes than residential. Moreover, this last fact appears to hold in suburban as well as central city areas.

In summarizing these descriptive results on land-use zoning near rail stations in southern California, three main points emerge. First, there is a good deal of variation by line. The patterns seem to follow from the geography of the urban area, as expected, with suburban communities having considerably more residential station areas than major employment centers. The second result is that station area "high-density" and "low-density" residential zoning patterns are different. Low-density suburban areas have the most low-density residential land near stations, while central city areas have the most high density. Third, for most lines, more land is devoted to residential uses near stations than to commercial uses—although the sum of commercial plus industrial nearly always exceeds residential.

Zoning Measure 2: Station Ratio

Looking simply at station shares for zoning categories has one major drawback. Some stations might have large shares in a particular use because the surrounding city also favors that use. The raw data certainly suggest that suburban municipalities do not have large amounts of high-density zoning near their stations. One obvious reason is because suburban municipalities do not have much high-density zoning anywhere in the city. The operative question is not whether a particular station has more or less commercial, residential, or other zoning, but how the zoning near stations compares to the rest of the city. If localities view station areas as attractive places for commercial uses, then there should be more commercial zoning near a station than elsewhere in the surrounding city. To measure how station area zoning patterns compare to zoning in the surrounding city, we constructed an additional zoning measure called the "station ratio."

The station ratio is the percentage of land within a quarter mile of the station in a particular zoning category divided by the percentage of land in the entire municipality in that category. For example, if 25 percent of the land within a quarter mile of a station is zoned residential while 50 percent of the land in that city is residential, the Station Ratio for that station is 0.5. In that case, the station area is half as residential as the jurisdiction as a whole. If the station ratio for a zoning category is less than one, the area near the station has a smaller share of its land zoned in that use than does the surrounding city. If the station ratio for a zoning category is greater than one, the area near the station has more of that land use (as a percentage of land area) than the surrounding city has.

Table 7.3 and figure 7.3 present the average station ratio for each rail transit line in southern California. First compare the ratios for residential and commercial land. Note that, for every rail line save one, the station ratio is larger for commercial zoning than for total residential. (The exception is the MTDB North line, which has a station ratio of 0.19 for commercial and 0.21 for residential.) Likewise, for every line save three, the station ratio for commercial zoning is larger than that for high-density residential. (The exceptions in this case are the MetroLink San Bernardino lines to Los Angeles and to Riverside and the MTDB East Line.)

Controlling for existing municipal zoning patterns, there is a stronger trend toward commercial than toward residential zoning near rail stations. This pattern is consistent across existing and proposed lines, lines in central and in suburban communities, lines that use heavy and light rail, and indeed essentially all lines in southern California. Also note that the station ratio is generally greater than one for commercial zoning, but often less than one for residential. This bolsters our claim that cities view stations more as sites for economic development than as residential locations.[8]

The good news for transit-based housing proponents is the large station ratio for mixed use zoning (see table 7.3). Yet most cities do not report mixed-use zoning for the municipality, and the station ratio for mixed use represents little other than the proposed lines in Orange County and north San Diego County. Also, the Station Ratio for mixed-use is relative to a very small base of mixed-use development in most mu-

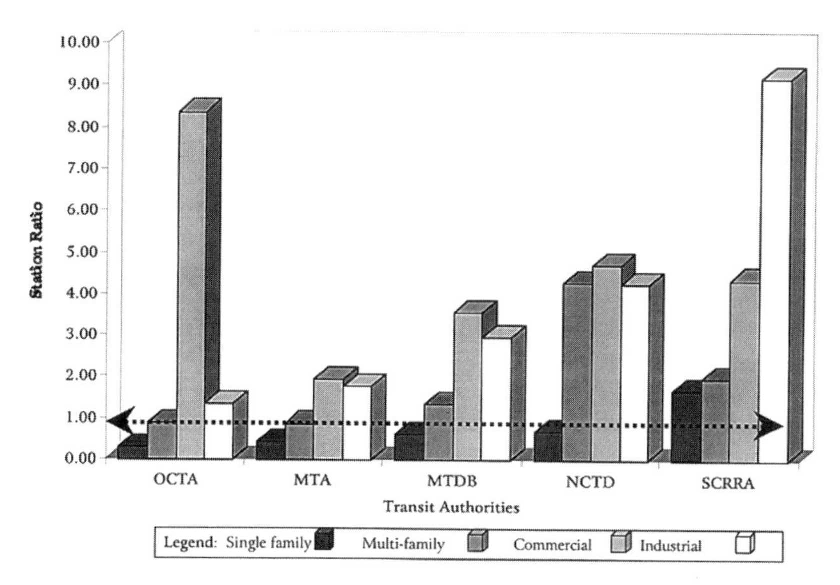

Figure 7.3. The ratio of station area zoning to city zoning, by transit authority ("station ratio," which equals 1 when the station area is zoned the same as the city).

nicipalities. Perhaps cities are receptive to mixed-use zoning near stations, but the clearer pattern is that cities are receptive to commercial, not residential, zoning near rail transit stations.

Interpreting the Evidence

Our suspicion is that the large values of station ratio for commercial zoning reflect municipal desires to use rail transit stations as centers of economic rather than residential development. This assumes either that municipalities adjust their zoning code once rail transit plans have been unveiled, or that municipalities exert some influence on the station siting process. Note that either could lead to the observed propensity toward commercial zoning near transit stations. The most likely explanation in our view is that municipalities exert some influence on the station siting decision.[9] For the purpose of inferring local incentives, however, it is unimportant whether the observed data are the result of zoning changes once stations are sited or of municipal influence on the station siting process.

There is also a third possibility that could give rise to the pattern discussed in the preceding section but that would give no information about municipal behavior. It is possible that a trend toward commercial zoning near stations reflects nothing more than the historical accident that southern California rail lines often used existing freight rail right-of-way.

Since industrial land uses are common and residential land uses somewhat rare near freight rail, it is possible that the existing rights-of-way used for rail transit were near land that was used primarily for business rather than residential purposes.[10]

Yet, for several reasons, we believe that historical right-of-way patterns are not driving the results described in the preceding section. First, the case of Orange County, which planned a line with no consideration of the existing right-of-way and sited stations in economic centers, is a clear counterexample. In particular, note that the station ratio data reported in table 7.3 show an especially pronounced trend toward commercial zoning near the stations on the OCTA main line (where the average station ratio for commercial is over 14).

Second, the consistency of the data in the preceding section argues strongly for a behavioral interpretation. While some lines were constrained by right-of-way, many were not. The fact that virtually all lines lean toward commercial development implies that something broader than right-of-way explains the trend. Third, we develop two behavioral models that predict land-use patterns near rail transit stations, and those models support our hypothesis that municipalities will want concentrations of commercial zoning near stations. That result holds when the models are fit only on data from stations on lines not constrained to use existing right-of-way, as shown in the following section.

Two Behavioral Models of Transit-Oriented Planning

The zoning data summarized above show that rail transit stations have, in an overwhelmingly large number of instances, more commercial zoning than does the surrounding city. Does that reflect municipal incentives to use land near stations for local economic or fiscal benefits? To answer that question, we developed two models of commercial zoning near stations. Both models test the hypothesis that the disproportionate share of commercial zoning near stations is due, in part, to local incentives. The first model focuses on municipal incentives to use stations to concentrate or enhance economic development within their city. The second model examines the more narrow hypothesis that zoning near stations reflects local fiscal pressures.

Model A: A Model of Zoning

If municipalities influence zoning patterns near rail transit stations, either through changes in the code or through station siting decisions, there should be systematic relationships between municipal characteristics and the observed zoning patterns. We tested for this first by developing a regression model of how municipal zoning behavior is linked to the ability to use rail transit to further local economic development goals.

In considering how economic development goals influence zoning near stations, it is important to realize that municipalities might specialize, with some choosing to become residential enclaves and others desiring to become employment growth centers. (See Tiebout [1956] for the seminal statement of this idea in the context of residential location choice.) We control for any desire to specialize by examining the station ratio for commercial zoning. The question studied below is not how much commercial zoning will be near stations in different cities, but rather how commercial zoning near stations compares to the rest of the city. Because the station ratio pegs zoning near stations to zoning in the surrounding city, the commercial station ratio controls for differences in local tastes regarding the amount of desired commercial zoning in the city.

The second important point to consider in constructing an empirical test is that all localities might want the same thing—more economic growth. Once we control for local tastes regarding commercial zoning, all municipalities should desire to compete equally for any benefits from economic development near rail stations. Observed differences in zoning behavior, then, are not due to different preferences about economic development (those are controlled through the commercial station ratio), but rather to differences in municipal abilities to act on a desire to use or site stations to focus commercial development. To test this idea, we use the regression model

$$\text{Station Ratio for commercial zoning} = \beta_0 + \beta_1 \text{ LINESHARE} + \beta_2 \text{ EMP}_{90\text{-}80} + \beta_3 \text{ DENSITY} + \beta_4 \text{ AREA} + \beta_5 \text{ POP90} + \beta_6 \text{ NSTATION} + u, \qquad (1)$$

where

LINESHARE = the number of stations on the line located in the municipality that contains that particular station, divided by the total number of stations on the line;

$\text{EMP}_{90\text{-}80}$ = employment change from 1980 to 1990 in the municipality that contains the station;

DENSITY = 1990 population divided by land area (land area measured in acres);

AREA = land area for the city that contains the station;

POP90 = 1990 population for the city that contains the station;

NSTATION = number of stations on the line;

u = the error term; and

β's = the parameters to be estimated.

The observations for the regression are the 232 existing and proposed stations in southern California as of 1995. The key independent variables are LINESHARE and $\text{EMP}_{90\text{-}80}$. Both capture how some localities might be better able to act on a desire to use stations as economic development centers. LINESHARE is intended to proxy for the amount of political influence that a locality can exert within the context of a rail line.

EMP_{90-80}, employment growth from 1980 to 1990, is intended to proxy for the extent that a municipality is well suited to economic development. We hypothesize that places with large amounts of growth during the 1980s are more likely to be able to attract economic development during the 1990s. If so, those high-employment-growth cities might be most likely to use rail stations to attract commercial development, and so we expect the sign on EMP_{90-80} to be positive.

The role of the LINESHARE variable is best explained by an example. Suppose a station is on a line with nine other stations. Also suppose the station is in a municipality that has three other stations (for the same line) within its borders. Thus, LINESHARE for that station (and all other stations on the same line within the same municipality) is 0.4 (the four stations within the city, divided by the ten total stations in the line). In general, LINESHARE must be greater than zero but less than or equal to one. Larger values of LINESHARE indicate that the station is within a municipality containing a larger fraction of the line's stations. Stated differently, larger values of LINESHARE show that the station is within a municipality that is "important" within the context of the rail line.

Presumably, municipalities with large portions of a line will have more influence over siting and coordinated land use near stations. Thus, the trend toward commercial zoning should be more pronounced in stations with large values of LINESHARE, and we expect the coefficient on LINESHARE to be positive in the regression equation.[11]

Population density in 1990 (DENSITY) was included in the regression because density is often thought to be linked to a city's land-use (and zoning) character.[12] We also included three variables that are correlated with LINESHARE and thus are expected to bias the coefficient on LINESHARE if omitted from the model. By definition, stations in large cities are more likely to have large values of LINESHARE (it is more likely that a large city is important in the context of any particular rail line). Likewise, stations on small rail lines are more likely to have large values of LINESHARE (it is more likely that any particular city can be important in the context of a small line). Thus, to be certain LINESHARE does not measure a large city or small line effect, we included the variables AREA, POP90, and NSTATION.

The regression results for all stations are reported in column 1 of table 7.4, and columns 2 and 3 show regression results for existing and proposed stations. Column 4 shows regression results for all stations outside of Los Angeles and San Diego. The results are rather consistent, and support our hypothesis that municipalities view stations as opportunities for economic and commercial development. The coefficient on EMP_{90-80} is significantly positive in columns 1–4, and LINESHARE is significantly positive in two of the first four columns. Yet the question remains whether the results demonstrate the effect of local incentives toward land near rail stations or whether the results simply reflect the historical legacy of zoning patterns near the often preexisting freight rail rights-of-way that were used for some of the rail transit lines in southern California.

Table 7.4: Economic Model of Station-Area Zoning[a]

Independent Variable	1 All Stations	2 Existing Stations	3 Proposed Stations	4 Exclude L.A. and S.D	5 Exclude Right-of-Way Constrained	6 Exclude SCRRA MetroLink	7 Exclude SCRRA and Right-of-Way Constrained	8 Exclude Orange County
LINESHARE	7.68*	4.17	13.48*	9.68	7.30	9.96*	24.80*	3.86*
	(2.74)	(1.78)	(2.34)	(1.67)	(1.40)	(2.62)	(2.96)	(2.30)
EMP_{90-80}	1.74×10^{-4}*	1.26×10^{-4}*	1.84×10^{-4}*	2.56×10^{-4}*	1.43×10^{-4}	1.85×10^{-4}*	1.22×10^{-4}	3.79×10^{-4}
	(3.75)	(2.12)	(2.44)	(4.31)	(1.86)	(2.79)	(1.03)	(1.04)
DENSITY	−0.88	−0.61	−0.93	−0.07	−1.21	−0.99	−1.39	−0.33
	(−5.75)	(−3.11)	(−4.15)	(−0.24)	(−4.95)	(−5.40)	(−4.87)	(−2.191)
AREA	-1.52×10^{-4}	-1.10×10^{-4}	-1.51×10^{-4}	2.29×10^{-4}	-1.33×10^{-4}	-1.63×10^{-4}	-1.10×10^{-4}	-4.53×10^{-4}
	(−4.50)	(−2.47)	(−2.67)	(1.91)	(−2.37)	(−3.55)	(−1.38)	(−1.65)
POP90	-5.15×10^{-6}	-3.71×10^{-6}	-6.21×10^{-6}	-6.08×10^{-5}	-4.79×10^{-6}	-5.82×10^{-6}	-8.20×10^{-6}	-1.59×10^{-6}
	(−4.82)	(−2.56)	(−3.61)	(−3.43)	(−2.61)	(−4.15)	(−3.41)	(−1.76)
NSTATION	0.10	−0.02	0.28	0.22	0.59	0.16	1.21	−0.04
	(1.45)	(−0.34)	(1.97)	(2.12)	(3.46)	(1.64)	(4.62)	(−1.02)
CONSTANT	8.68	8.95	4.05	−1.79	6.22	8.06	−7.07	6.77
	(4.72)	(5.28)	(1.00)	(−0.48)	(1.91)	(2.79)	(−1.15)	(5.82)
R^2	0.20	0.20	0.21	0.24	0.29	0.22	0.42	0.42
R^2 adj.	0.17	0.15	0.17	0.21	0.25	0.19	0.37	0.37
N	211	101	110	145	107	164	79	79

[a]The share of station area land zoned commercial, relative to the share of city land zoned commercial. *T* statistics are shown in parentheses.

*Those coefficients on LINESHARE and EMP_{90-80} that are statistically significant at the 5 percent level (two-tailed test).

The problem with right-of-way constraints is best handled by omitting stations on lines that were constrained to use existing rights-of-way.

We used two techniques to identify lines that were constrained to use existing rights-of-way.[13] Both methods were based on excluding lines with a large amount of industrial zoning near stations, on the assumption that heavy concentrations of industrial zoning are a sign of the land-use legacy of preexisting freight rail routes. By identifying stations with high concentrations of industrial zoning (specifically, those stations with industrial station ratios in the 80th percentile or above), we excluded the following seven lines: MetroLink Riverside to Los Angeles, MetroLink San Bernardino to Los Angeles, MTA Blue, MTA Green, NCTD Oceanside to San Diego, MTDB South, and MTDB Centre City.[14] The results of excluding those seven lines from the regression in equation (1) are reported in column 5 of table 7.4. We also omitted all SCRRA MetroLink stations from the regression, again based on the fact that those lines had heavy nearby concentrations of industrial zoning at several stations. The regression results are reported in column 6. In column 7 we exclude all stations on the lines that were identified as right-of-way constrained by either of the two criteria used in columns 5 and 6.[15]

The results in columns 5–7 of table 7.4 continue to support the hypothesis that high commercial station ratios reflect local incentives to use rail stations to enhance or attract economic development. Two of the three models in columns 5–7 yield significant coefficients for at least one of the LINESHARE and $EMP_{90\text{-}80}$ variables. Note further that while neither LINESHARE nor $EMP_{90\text{-}80}$ are significant at the 5 percent level in column 5, $EMP_{90\text{-}80}$ is significant using a 10 percent two-tailed test. In the last column of table 7.4, we exclude all stations in Orange County, to show that the significant coefficient on LINESHARE is not simply due to the influence of those stations.

Overall, the results in table 7.4 are consistent with the descriptive data presented above, and strongly support our argument. The city employment growth rates and shares of the transit system are in most cases positively and significantly associated with commercial zoning near rail transit stations, controlling not only for the land-use character of that community, but also for other relevant community characteristics as well. Growing employment centers use station areas for commercial development. In addition, the more influence a city has over station siting and coordination, as measured by the share of the system within jurisdiction borders, the more likely it is that a municipality will concentrate economic development activities in the station areas. These patterns hold up across virtually all city and rail line types.

Taken together, both facts suggest that the observed trend toward commercial zoning near stations reflects municipal intentions. At least one of the two coefficients (LINESHARE or $EMP_{90\text{-}80}$) is significant in all regressions but one, and the tendency toward commercial zoning appears somewhat insensitive to the nature of the rail lines, the right-of-way used, or the size or character of the cities containing these lines. This pattern supports our hypothesis that municipalities tend to view stations as centers of economic development.

Model B: Zoning and Public Finances

A further test of municipal incentives toward zoning near rail stations focuses specifically on the role of fiscal pressures. The relative concentration of commercial zoning near rail transit stations is consistent with our hypothesis that municipalities see stations as opportunities to enhance their fiscal position. In this section, we test that hypothesis.

Assigning each station to the city, or sphere of influence, in which it is located gives us 65 municipal jurisdictions in the data set, spanning six counties.[16] These represent a diverse array of large central cities, emerging suburban centers, and small communities. Various fiscal data were collected for these communities, and several measures of fiscal character were in turn calculated. The raw data included sales tax collections, property tax collections, total revenues, and total expenditures for the fiscal years 1980/81 through 1990/91. These were then put into per-resident terms for each year and adjusted for inflation, and the rates of change over the period were determined. In addition to having total tax collections and per-capita collections for both the sales and property tax, as well as the percentage real change in those numbers, the share of each tax's contribution to total revenues was also calculated as a measure of the community's "dependence" on each tax base. Tax base levels and trends were also examined.

Several of these aggregate revenue patterns are presented in table 7.5. The average city saw its property tax collections increase twice as fast as sales tax collections over the 1980s, to the point that they comprised about a quarter of the average city revenue pie at the end of that period. This partly reflects the low property tax collections in 1980/81 resulting from Proposition 13 rolling back both property tax rates and property as-

Table 7.5: City Revenue Levels and Trends

	1991/92 Levels		% Real Change, 1980/81–1991/92	
Revenue Measure	City Average	Standard Deviation	City Average	Standard Deviation
Property taxes	$17,304,402	(71,108)	87.9%	(71.8)
Sales taxes	$13,927,614	(33,621)	38.0%	(80.5)
Property taxes per capita	$87	(51)	46.8%	(89.7)
Sales taxes per capita	$128	(150)	0.5%	(59.6)
Share of property taxes in general revenues	24.5%	(12.2)	111.8%	(132.8)
Share of sales taxes in general revenues	30.7%	(13.4)	33.8%	(42.7)

sessments substantially. The property tax base then steadily grew through the California housing market expansion of the 1980s. Sales tax collections grew but more slowly, especially as a share of all revenues. The average per-capita growth of sales tax collections was flat. Both taxes increased their importance in local revenues, but the property tax by more. Again, this reflects the small property tax base left in the wake of Proposition 13. Thus, the main source of fiscal growth through this period was population growth and associated land development.

The dependent variable for the regression model is again the station ratio for commercial zoning. The independent variables are intended to accomplish two things. First, they measure fiscal characteristics that might help explain the observed tendency toward relatively large commercial station ratios reported in table 7.3. Second, the independent variables help control for other, nonfiscal, determinants of zoning that might be correlated with the fiscal effects that we are testing for here.

The fiscal independent variables measure the extent to which the municipality that contains the station depends on local sales and property tax revenues, controlling for total budget size. If the tendency toward high commercial station ratios is due in part to fiscal influences on zoning and siting, then we hypothesize that the commercial station ratio will be larger in those municipalities that depend more heavily on either sales or property tax revenues. Those variables are labeled STAXDEP, for sales tax dependence, and PTAXDEP, for property tax dependence. Between the two, we expect there to be a stronger link between sales tax dependence and station-oriented commercial zoning because commercial development can generate taxable sales. The property tax has become more important in recent years, relative to 1979, as property values rose and new property was developed to accommodate a much larger population, but this is not due exclusively to commercial development. To control for the size of the local budget in per-capita terms, we also include current municipal expenditures per capita.

The nonfiscal variables included in the model are population density, median household income, and percentage of the population that was White in the municipality that contains the station. All those variables are measured in 1990, and all control for nonfiscal factors that might influence development patterns near stations. Because the region's rail lines run through cities that span the gamut from large urban centers to small outlying communities, we include density to control for the development differences that one would naturally expect with such variation.[17] Income and percent White are intended to proxy for "municipal tastes" toward land use policy that might also be correlated with the tax instruments we focus on here.

The public finance regression model is thus

$$\text{Station ratio for commercial zoning} = \alpha_0 + \alpha_1\ \text{STAXDEP} + \alpha_2\ \text{PTAXDEP} + \alpha_3\ \text{DENSITY} + \alpha_4\ \text{INCOME} + \alpha_5\ \text{EXPCAP} + \alpha_6\ \text{PCTWHITE} + e, \quad (2)$$

where

STAXDEP = fiscal year 1990/91 sales tax revenues divided by total municipal revenues,

PTAXDEP = 1990/91 property tax revenues divided by total municipal revenues,

DENSITY = 1990 population divided by land area (land area measured in acres),

INCOME = 1990 median income,

EXPCAP = 1990/91 municipal expenditures divided by 1990 population,

PCTWHITE = proportion of 1990 population that is White,

e = the error term, and

α's = the parameters to be estimated.

Note that, for each station, the independent variables measure characteristics of the city that contains the station. That allows considerable variation since there are 65 municipalities containing stations in southern California. (Later in this section, we also show that the results are not proxying for the effects of the region's two largest cities, Los Angeles and San Diego.) Note also that the two tax dependence variables, STAXDEP and PTAXDEP, are not collinear. For most cities in our data set, those two tax instruments provide much less than the total revenue in the municipality. On average, sales and property taxes provide 56 percent of local revenues for the cities in our dataset. Only three cities get more than 75 percent of their tax revenues from sales and property taxes combined.[18] Thus, a greater reliance on one tax instrument (e.g., the sales tax) does not necessarily imply a reduced reliance on the other (e.g., the property tax).

The results of fitting the model in equation (2) on data for all existing and proposed southern California stations are shown in model (a) of Table 7.6.[19] Note that the coefficient on STAXDEP is significantly positive using a 5 percent one-tailed test. The other significant variables are density, income, and percent White. The positive coefficient on income could reflect the ability of high income cities to support commercial centers.

The negative coefficient on DENSITY suggests that, in low-density cities, it is easier for station area commercial concentrations to be more different (i.e., more commercial) than the rest of the municipality. The negative coefficient on PCTWHITE is more difficult to explain, but interpreted literally municipalities with smaller minority populations have lower commercial station ratios. For now, note that our attention is focused on the fiscal variables, and the other independent variables, PCTWHITE included, are in the model largely to be certain that STAXDEP and PTAXDEP do not proxy for other, nonfiscal, influences.

In model (b), we add a dummy variable that equals one if the station has not yet opened as of 1996. This is intended to control for differences between existing and proposed stations, and can at least partially illu-

Table 7.6: Fiscal Model of Transit-Area Zoning: Dependent Variable Is Station Ratio for Commercial Zoning

Independent Variables	Model A Does ratio of station area commercial zoning to city average depend on revenue mix, controlling for density, race, and household income?	Model B Also controlling for station status	Model C Same as Model B but excluding LA and San Diego observations	Model D Same as Model B but excluding Orange County observations	Model E Same as Model B but controlling for "industrial ratio" right-of-way constrained stations
CONSTANT	−5.15	−5.23	−5.11	0.99	−4.57
	(1.21)	(1.23)	(0.96)	(0.33)	(1.04)
Share of revenues from sales taxes in 1990/91 (STAXDEP)	9.96*	10.65*	12.04*	5.47	10.82*
	(1.87)	(1.99)	(1.84)	(1.51)	(2.01)
Share of revenues from property taxes in 1990/91 (PTAXDEP)	6.14	7.42	8.67	11.68*	6.60
	(0.89)	(1.06)	(1.04)	(2.89)	(0.93)
Current expenditures per capita in 1990/91 (EXPCAP)	1.60×10^{-4}	1.80×10^{-4}	1.65×10^{-4}	1.63×10^{-4}	1.79×10^{-4}
	(0.87)	(0.97)	(0.76)	(1.59)	(0.96)
1990 population density	−0.49*	−0.53*	−0.58*	−0.36*	−0.54*
	(2.95)	(3.10)	(2.84)	(3.48)	(3.13)
1990 median household income	3.60×10^{-4}*	3.41×10^{-4}*	3.32×10^{-4}*	3.59×10^{-5}	3.28×10^{-4}*
	(4.76)	(4.42)	(3.64)	(0.67)	(4.02)
Share of 1990 population that is White	−0.07*	−0.08*	−0.76	−0.02	−0.08*
	(1.77)	(1.88)	(1.49)	(0.77)	(1.80)
Dummy = 1 if station not in operation		1.22	0.97	0.56	1.32
		(1.13)	(0.62)	(0.86)	(1.12)
"Industrial" station ratio					−0.02
					(0.62)
R^2	0.20	0.20	0.19	0.14	0.20
R^2adjusted	0.17	0.17	0.15	0.10	0.17
N	214	214	150	170	212

The absolute values of the t-statistics are shown in parentheses.

minate the relative role of zoning changes versus siting in any fiscal behavior toward stations.

Yet we caution that commercial concentrations near existing stations might reflect past siting decisions, and commercial concentrations near proposed stations might be a function of zoning changes that have been made in anticipation of the station opening. The purpose of the dummy variable is thus not to definitively separate zoning changes from siting decisions, but to ensure that the relationship between STAXDEP and the commercial station ratio persists when we control for the existing/proposed status of stations. The results are essentially unchanged from those in model (a), and the coefficient on the dummy variable for station operating status is not significant at the 5 percent level.

Model (c) of table 7.6 excludes the stations in Los Angeles and San Diego. The only change is that the coefficient on PCTWHITE is now insignificant using a 5 percent one-tailed test. This suggests that this variable was partially proxying for the effect of the two large central cities in the region, which have larger minority populations than many of the more outlying communities. Yet note that even when Los Angeles and San Diego are omitted from the analysis, the coefficient on STAXDEP is still significantly positive.

Model (d) excludes all stations in Orange County. Orange County stations are excluded since those showed an especially pronounced tendency toward commercial. (Recall that the station ratio for commercial for all OCTA stations was 8.4.) This is the only specification in which STAXDEP is not significantly positive. But note that PTAXDEP is significantly positive in model (d), giving some evidence that local fiscal policy affects zoning near stations. Other changes in model (d) are that the coefficients on INCOME and PCTWHITE are both insignificant. This suggests that, in models (a)–(c), the income variable was to some extent proxying for the differences between Orange County's system and the other lines in the region.

The stations on the proposed OCTA lines were specifically designed to link commercial and employment centers in the county (Central Orange County Fixed Guideway Project, 1990). It is not surprising that some of the link between sales tax dependence and commercial station ratios is due to the influence of the Orange County stations in the regression analysis. Yet it is notable that, when the Orange County stations are omitted from the regression analysis, the coefficient on PTAXDEP becomes statistically significant. The link between fiscal variables and commercial station ratios persists when Orange County stations are omitted, but the nature of that link changes.

This link highlights two important points. First, the fiscal behavior we examine here varies across municipalities. While we hypothesized that all municipalities have incentives to use rail transit stations to enhance their fiscal positions, certainly some cities will be better able to act on those incentives. The cities along the Orange County line were intimately involved in the early station siting decisions, and that might be one reason why the link between STAXDEP and commercial station ratios is influenced by the Orange County system.

Second, the importance of various tax instruments also varies across cities. Excluding Orange County, there is a positive link between PTAXDEP and commercial station ratios, suggesting that the property tax is more of a focus for municipal fiscal behavior outside of Orange County. Because Orange County has recently become home to some of the region's newer employment and commercial centers, the taxable sales that those centers generate might be more visible to municipal leaders in that county. Elsewhere, the property tax appears to be an important source for fiscally motivated planning decisions.

We also tested these results to examine whether the link between the fiscal variables and commercial station ratios reflects the legacy of freight rail right-of-way rather than municipal incentives. In the context of the public finance regression model in table 7.6, the question is whether cities with freight rail developed commercial concentrations that led those cities to become dependent on the sales tax. If those commercial concentrations were also near freight rail lines, current rail transit stations might have large commercial station ratios. In short, sales tax dependence and commercial station ratios could be positively related not for fiscal reasons, but because each are influenced by the land-use legacy of freight rail lines in the region. Similarly, if commercial concentrations near old freight rail lines generate large property tax revenues, the link between property tax dependence and commercial station ratios could reflect the influence of preexisting rail right-of-way. We conducted two tests to examine this possibility.

First, we identified lines that were constrained to use existing rail right-of-way, using the definitions of right-of-way constrained lines described above. For each definition, we examined whether the cities along right-of-way constrained lines are more dependent on either the sales or the property tax when compared to other cities with rail stations. Such a pattern would suggest the possibility that the fiscal and station ratio variables both reflect the influence of preexisting land use patterns, rather than local incentives for land use near or the siting of rail stations.

For all three definitions of right-of-way constrained lines, the only differences in property and sales tax dependence were counter to the pattern that one would expect if freight rail created a dependence on either tax instrument. Cities with stations on right-of-way constrained lines were either no different from other cities in terms of the proportion of revenues generated from property and sales taxes, or cities on right-of-way constrained lines were less dependent than other cities on those tax instruments.[20] This provides some assurance that the relationship between sales tax dependence, property tax dependence, and commercial concentrations near stations is not due to the legacy of freight rail in the region.

The second test is a more explicit attempt to control for the possible impact of freight rail right-of-way on the land use character near each station. If freight rail lines are near predominantly industrial and commercial uses, we can control for that influence by including a measure of industrial land use in the regression analysis.

Model (e) of table 7.6 reports the results of including the station ratio for industrial zoning in the regression reported in model (b). If the relationship between STAXDEP and commercial station ratios is simply an artifact of land use patterns associated with preexisting freight right-of-way, controlling for such uses by including a measure of industrial land use should cause the coefficient on STAXDEP to become insignificant. Yet STAXDEP is still statistically significant in model (e), and the coefficients on the other variables in the model are essentially the same as in model (b).

Overall, cities with lines that were constrained by freight right-of-way do not have any greater reliance on either the sales or the property tax, and explicitly controlling for station-area industrial zoning does not change the results reported in model (b) of table 7.6. This suggests that the positive association between STAXDEP and the commercial station ratio is not due to any historical concentration of industrial and commercial land near transit lines that follow freight rail right-of-ways. Instead, the regressions in table 7.6 provide evidence that fiscal motives influenced either the siting of stations, the zoning near stations, or both in ways that lead rail transit stations in southern California to have considerably more commercial land than the communities where they are located.

Many of these new types of fiscally motivated planning behaviors have not been studied in great detail. While TOD is one context in which to study the fiscal influence on land-use planning, it certainly is not the only opportunity. Earlier literatures studied incentives for fiscal zoning and attempts to increase the local tax base, but those literatures typically focused on the property tax (e.g., Mills and Oates, 1975). In California, and likely in other states also, fiscal pressures are increasingly focusing on land uses that generate sales tax revenue. Fiscal competition now is over commercial uses, and the ramification of these new fiscal pressures are not fully understood. Future research should study other planning behaviors to see if the fiscal pressures associated with transit area planning are also typical of other aspects of land use planning.

Summary

The morals of this chapter are that the literature has inadequately accounted for the motivations of local governments toward transit-based housing, and that these motivations matter. Our results indicate a conflict between local and regional rail transit goals, owing mainly to a mismatch in the distribution of the costs and benefits of TOD. The advantages of transit-based housing, such as increased ridership, accrue largely to the region. The advantages of transit-area commercial developments accrue disproportionately to those local governments that reap their fiscal advantages. This suggests that municipal land-use choices may not exploit the regional advantages of rail transit—because they were never intended to.

If this argument sounds similar to the jobs-housing balance debate, it should (e.g., Cervero, 1989b; Giuliano, 1992). In both debates, local land-use patterns are perceived to create an "imbalance" from a regional transportation perspective. The potential imbalance in our case is in the form of an excessive number of employment and shopping "destination" stations relative to the number of residential "origin" stations. That is, left to their own devices, almost every city wants the train to bring people into town in the morning rather than send them elsewhere. The end result may be too few communities acting as feeders to sustain the health of the larger rail system, especially as employment continues to decentralize to the suburbs.

This brings up two points mentioned in chapter 6. First, the barriers to TOD, and more generally to transportation plans that include urban design elements have possibly been underestimated due to an insufficient focus on local incentives toward land use. Second, the argument that local municipalities will have incentives to restrict the supply of certain types of land uses has some support in the case of transit-based housing. This has implications both for the broad approach of using urban design to bolster transportation policy and the more narrow issue of TOD.

Regarding TOD, transit-oriented plans have to appeal to municipalities' self interest to be successful. Elsewhere, we suggested that the most promising way to do this is to encourage localities to identify situations in which residential TOD can bring the same economic benefits normally associated with commercial and office concentrations (Boarnet and Crane, 1997). Local governments can be, and often already are, educated to recognize those instances where building large residential tracts provides a ready demand for local goods and services. In such cases residential development, because it supports commercial retail development, in turn generates substantial local economic benefits.

To the extent that this happens near rail transit stations, mixed-use TOD plans might be the best way for a city to reap fiscal and economic benefits from their land. Rather than merely enabling localities to build transit-based housing, regional authorities should actively identify those situations where cities can benefit from a TOD that includes housing. A notable example is the new Los Angeles General Plan (City of Los Angeles, 1993), which includes residential development as a secondary element around stations that are often commercial nodes. The increased residential development near those stations is viewed as a way of both increasing ridership and enhancing the economic health of the commercial development in the area.

More generally, municipalities certainly appear to plan with local benefits foremost in their mind, and at least in the case of TOD this can conflict with broader regional plans. This bolsters the argument that local governments might restrict the supply of some urban designs. Most important, the evidence in this chapter suggests that understanding local incentives toward land use and neighborhood design is vital.

Almost all of the discussion about the transportation benefits of the new urban designs has focused on their impact on travel behavior, leaving the question of local incentives and the municipal regulatory environment all but untouched. In the case of TOD, this is an important oversight, and we suggest that local incentives can be similarly important with respect to the new urban designs more generally. For that reason, in chapter 8 we continue with an in-depth case study of local attitudes and behavior toward TOD implementation.

In this chapter we examine the experience of one urban area, San Diego County, California, to assess the implementation of transit-oriented development projects along the oldest of the current generation of light rail transit lines in the United States. By analyzing local zoning codes, planning documents, and detailed interviews with planning directors, we illustrate both the opportunities for and barriers to transit-oriented projects in the region. We further characterize the transit station area development process in San Diego County and draw insights into the prospects for implementation in other urban areas.

8 A Case Study of Planning

The facts, figures, and inferences in chapter 7 regarding municipal behavior toward transit-oriented housing opportunities illustrate many points. Still, there is much that even a careful statistical analysis might miss or misunderstand. For that reason, we also explored what we could learn by talking to real planners about these issues.[1]

The case of San Diego is interesting and useful for several reasons. First, the San Diego Trolley is the oldest of the current generation of light rail projects in the United States. Unlike many newer systems, the age of San Diego's rail transit (the South Line opened in 1981) allows time for land use planning to respond to the fixed investment. Second, the San Diego system is no stranger to modern transit-based planning ideas. The San Diego City Council approved a land-use plan for their stations that includes many of the ideas promoted by transit-oriented development (TOD) advocates (City of San Diego, 1992). Third, the light rail transit (LRT) authority in San Diego County, the Metropolitan Transit Development Board (MTDB), is often regarded as one of the more successful municipal LRT agencies. The initial parts of the MTDB rail transit system were constructed strictly with state and local funds, using readily available, relatively low-cost technology (Demoro and Harder, 1989, p. 6). Portions of San Diego's system have high fare-box recovery rates, including the South Line, which in its early years recovered as much as 90 percent of operating costs at the fare box (Gómez-Ibáñez, 1985).

All of these factors make San Diego potentially a "best-case" example of TOD implementation. When generalizing from this case study, it is important to remember that the transit station area development process in San Diego is likely better developed than in many other urban areas in the United States. The results from San Diego County can illustrate general issues that, if they have not already been encountered, might soon become important in other urban areas with rail transit systems. Also, given San Diego County's longer history of both LRT and TOD when compared with most other regions, any barriers identified in San Diego County might be even more important elsewhere.

Data Sources and Background

We focused our attention on three different sources of information. First, we gathered data on land use near San Diego Trolley stations. This included zoning data for all land within a quarter mile of each station. One of the co-authors also visually inspected the development near every San Diego Trolley station. Second, we obtained general and specific plans, minutes of local planning board and city council meetings, and local news articles.

The third information source was a series of detailed interviews with local planning directors in each of the seven cities with one or more stations on existing San Diego Trolley lines as of 1995. We focused on planners because our goal was to understand the role of planning in the TOD implementation process. While previous studies have examined the roles and attitudes of developers (Bernick, 1990), residents (Cervero and Bosselmann, 1994), and transit authorities (Cervero, Bernick, and Gilbert, 1994, esp. pp. 8–10) toward TOD, planners have been relatively overlooked.

Interviews with Planning Directors

The outline for the interviews was developed in early 1995. That outline was pretested on city planners in two jurisdictions outside of the study area, and then updated by clarifying potentially confusing questions. The initial interviews for this study were conducted in August and September of 1995. The planning directors in each of the seven San Diego County cities with rail transit stations agreed to be interviewed.[2] All had at least thirteen years of experience in the planning profession, and each had been with their city at least six years except in San Diego, where the planning director had been with the city for two years as of 1995.

Respondents were asked a series of questions, over the course of about an hour, designed to illuminate their city's goals for development near its rail transit stations, the steps taken toward those goals, and any opportunities or barriers in the development process. To avoid eliciting opinions that deferred to recent writings on the topic, the term "transit-oriented development" was not used at any time prior to or during the interview.[3] The format was open-ended; the respondents were allowed to elaborate on each question as they saw fit. Each interview was taped using an audio-cassette recorder, and the interviews were transcribed verbatim. Following transcription, responses were analyzed both for their uniqueness and for general patterns.

Because the initial results suggested that the City of La Mesa played a special role in San Diego County TOD, we returned to that city for a second interview in April of 1997. Our primary objective for the second interview was to verify information on the planning process near La Mesa's rail transit stations during the 1980s. We included both Dave Witt, currently the planning director in La Mesa, and Dave Wear, currently La Mesa's City Manager, in the second interview. Wear had been Planning

Director in La Mesa during the 1980s, and Witt was Assistant Planning Director during the 1980s. We followed the same procedure in the second interview as in the first interview. Our general conclusions were reinforced by the second interview, and we uncovered nothing at odds either with the first interview or with other information. Overall, we do not believe that the time gap between the first and second interviews influenced either the interviewees' recollection of the events or their willingness to provide information.

Given the small number of cities (seven) and the unique characteristics of each station and nearby projects, we considered it unrealistic to believe that the interview subjects could remain anonymous. For that reason, we informed each participant before the interview of the possibility that the study results would be disseminated in working paper or published form, and of the possibility that they would be identified. While that was necessary, it does raise questions about whether the subjects' responses were influenced by their knowledge that the results would be disseminated. To minimize this concern, we used archival research when possible to cross-check the planning directors' comments with other accounts of the same process.

The San Diego Trolley

The development of the San Diego Trolley began with legislation introduced into the California State Senate by James R. Mills in 1975. Mills's bill required that a percentage of highway funds be allocated to rail projects in Los Angeles, Orange, and San Diego counties. After Los Angeles and Orange counties objected to certain provisions in the bill, they were dropped from the legislation. The bill, which then applied only to San Diego, contained two major stipulations. First, funds must be spent within five years or the money would be returned to the state. Second, only off-the-shelf technology already operating successfully elsewhere could be selected for the rail transit project (Demoro and Harder, 1989, p. 6).

Once the legislation passed, and the San Diego MTDB was created to implement the rail plan, there was little time to choose a route and build the project. An established freight rail route to the north of the city, part of the San Diego and Arizona Eastern Railroad (SD&AE), was the logical choice, since that was where most new development in the county was occurring. Then in 1976 Hurricane Kathleen washed out major sections of the SD&AE. The MTDB settled for a southern route, beginning in downtown and terminating in San Ysidro, near Tijuana and the Mexican border (Demoro and Harder, 1989, p. 6). On July 26, 1981, the San Diego Trolley began service on the South Line. Revenue service on the initial 4.5 mile segment of the Trolley's East Line, to Euclid Avenue, began in March of 1986, and extensions to the East Line opened in 1989, 1990, and 1995. The initial portion of the Trolley's North Line, a 3.2-mile extension from County Center/Little Italy to Taylor Street in Old Town, opened in 1996.[4] The Mission Valley Line, which roughly parallels the

8 Freeway to the north of downtown, opened in late 1997, after the research for this case study was concluded.[5] The MTDB rail system is shown in figure 8.1.

Zoning and Land Use Patterns Near Trolley Stations

A good starting point for assessing both potential and actual TOD in San Diego County is to summarize zoning and land-use patterns near existing trolley stations. Calthorpe (1993, p. 63) recommends that TODs have at least 20 percent of their land area devoted to housing, and that criterion is also specified in the City of San Diego's (1992) transit-oriented

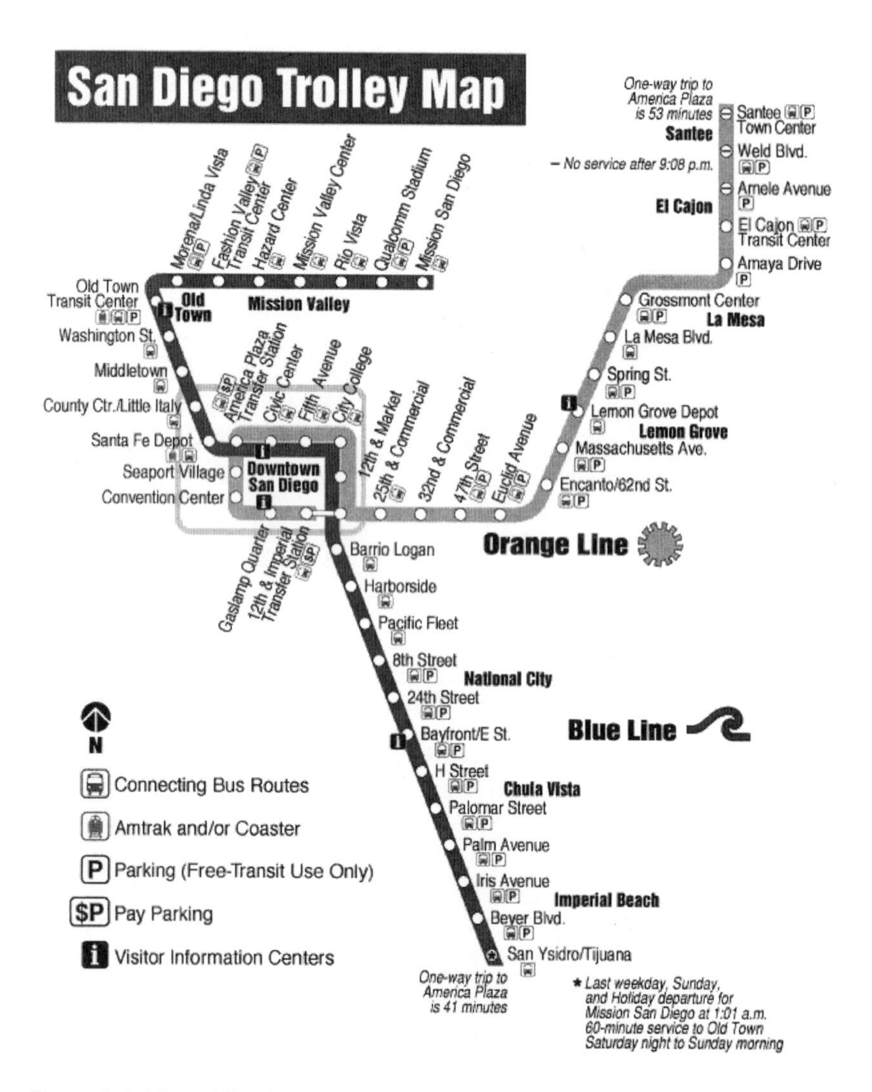

Figure 8.1. Map of San Diego trolley system.

guidelines. According to Calthorpe (1993, pp. 64, 83), average residential densities within TODs should be at least ten dwelling units per acre for neighborhood TODs and at least fifteen dwelling units per acre for more centrally located, or urban, TODs. Calthorpe (1993, p. 78) also suggests minimum floor area ratios (FARs) of 0.3 for retail with surface parking and 0.35 for offices without structured parking, but he encourages higher FARs for both types of development.

With those guidelines in mind, zoning data were gathered for land within a quarter mile of the forty-eight trolley stations that were open or under construction as of the summer of 1995.[6] Table 8.1 reveals that nineteen of the forty-eight stations have more than 20 percent of their nearby (quarter-mile radius) land zoned residential.[7] The "multifamily residential" column corresponds, for the most part, to a minimum density of fifteen dwelling units per acre.[8] This is the density cutoff suggested by Calthorpe (1993) for what he calls "urban TODs," although it is slightly higher than the suggested minimum density in Calthorpe's "neighborhood TODs." Overall, the Multifamily column in table 8.1 can serve as a good approximation to the density levels compatible with what is commonly discussed in transit-based housing plans. When focusing only on multifamily residential, the scope for transit-oriented housing drops dramatically. Only eight stations in San Diego County have more than 20 percent of their nearby land zoned at residential densities (Multifamily in table 8.1) compatible with the "urban TOD" definitions in Calthorpe (1993) and the City of San Diego (1992) guidelines.

Commercial and mixed-use zoning predominates along the Centre City Line in downtown San Diego, and commercial zoning is also found near several stations on other lines. Comparing allowable FARs across municipalities is rather difficult. Some cities do not specify FAR requirements, instead focusing on parking and height requirements. Thus, it is not possible to develop a FAR criterion that approximates some standard of TOD commercial zoning in the way that "multifamily residential" approximated the suggested densities for residential TOD.

Instead, we gathered information on the dominant land uses near each station. This is from a visual inspection of all station areas. Those dominant nearby uses are shown to the right of the zoning data in table 8.1. Note that forty-three of the forty-eight stations have dominant nearby uses that include a commercial or industrial component. This reflects a pronounced tendency toward commercial and industrial land uses near San Diego Trolley stations, consistent with our previous studies of southern California rail transit (Boarnet and Crane, 1997, 1998a).

The information in table 8.1 suggests that, in many instances, local zoning and existing land uses are not consistent with transit-oriented housing. Yet to make more specific statements about TOD implementation in San Diego County, it is necessary to examine specific projects.

In table 8.2, we list all of the San Diego County projects, both existing and under construction, that we judged to be consistent with the TOD definition outlined above. We used a two-step process to identify projects

Table 8.1: Zoning Within a Quarter Mile of San Diego Trolley Stations, with Dominant Nearby Land Uses

Municipality or Jurisdiction	Line/Station	Zoning Categories				Dominant Nearby Land Use
		Multi-family	Total Res.	Commercial	Mixed	
San Diego City	North					
	Taylor St. (Old Town Sta.)	0%	0%	0%	2.90%	Presidio Park—Old Town S.D.
	Washington	0%	3.40%	22.40%	6.50%	Industrial—S.D. Intl. Airport/US Marine Corps. (MCRD)
	Airport and Palm	0%	12.60%	0%	3.60%	Industrial—S.D. Intl. Airport
	Mission Valley East					
	Morena Area	0%	<1%	0%	0%	Industrial—near junction I-8 and I-5
	Fashion Valley	0%	0%	74%	8%	Auto-oriented commercial—Fashion Valley Mall
	Hazard Center	28.40%	28.40%	37.60%	0%	Auto-oriented commercial/office/hotel—Hazard Center
	Mission Valley Center	9%	27.40%	34.50%	0%	Auto-oriented commercial—Mission Valley Shopping Ctr.
	Rio Vista Center	13.30%	17.90%	30.60%	7.70%	Undeveloped/retail/office—Rio Vista West
	Jack Murphy Stadium	0%	0%	25.90%	0%	Recreation/commercial—Jack Murphy Stadium
	Rancho Mission	39.70%	42.80%	<1%	0%	Multifamily residences—near junction I-8 and I-15
	Centre City					
	County Center/Little Italy	0%	0%	23%	0%	Office/commercial—S.D. County Admin. Ctr.
	Santa Fe Depot	0%	3.60%	24.10%	5.10%	Office/commercial—Santa Fe Depot
	Seaport Village	0%	3.60%	3%	20.40%	Office/commercial/multifamily residences
	Convention Center West	0%	0%	0%	28.10%	Hotels/multifamily residences—S.D. Convention Center
	Gaslamp Convention Center	0%	0%	0%	12.50%	San Diego Convention Center
	American Plaza Transfer Sta.	0%	3.6%	30.30%	2.10%	Office/commercial/rail stations
	Civic Center	0%	0%	59.90%	2%	Government offices/jail/hotel—Civic Center
	Fifth Avenue	0%	0%	57.40%	7.20%	Office/multifamily res./parking
	City College	2%	2%	39.90%	0.80%	Institutional/commercial—S.D. City College
	Market & 12th	0%	0%	3.40%	4.3%	Industrial/commercial
	Imperial & 12th Transfer Sta.	0%	0%	0%	17.30%	Commercial office/light industrial—MTS Rail Yard
	South					
	Barrio Logan	0%	0%	0%	63.60%	Industrial/multifamily res.—Coronado Toll Bridge

San Diego/ Naval Reserve	Harborside	0%	0%	0%	19.70%	Heavy industrial/commercial—National City Steel Works
Naval Reserve	Pacific Fleet	0%	0%	0%	0%	Heavy industrial—National City Steel Works
Naval Reserve/ National City	8th St	0%	0%	0.60%	0%	Commercial/industrial—I-5 corridor
National City	24th St.	0%	0%	10.20%	0%	Auto-oriented commercial—I-5 corridor
Chula Vista	Bayfront/E. St.	15.30%	15.30%	48.20%	0%	Auto-oriented commercial—I-5 corridor
	H. St.	30%	34.40%	33%	0%	Commercial/mobile home park—I-5 corridor
	Palomar St.	2.90%	25.80%	33%	0%	Auto-oriented commercial—I-5 corridor—Palomar Center
San Diego City	Palm Ave.	0%	55.60%	0%	7%	Commercial/single-family res./mobile home park
	Iris Ave.	0%	41.60%	0%	1.80%	Multifamily res./industrial
	Beyer Blvd.	0%	53.40%	10.30%	0%	Multifamily residences
	San Ysidro/International Border	0%	0%	10.50%	0%	Border commercial—US/Mexico Border
	East					
	25th & Commercial	43.70%	43.70%	0%	13.60%	Light industrial/commercial/single-family res.
	32nd & Commercial	51.20%	51.20%	0%	6.30%	Heavy industrial/single-family res.
	47th St.	44.80%	45%	0%	7.60%	Multifamily res./commercial—Creekside Villas Apts.
	Euclid Ave.	16.20%	28.50%	8.20%	0.60%	Single-family res./commercial
	Encanto/62nd St.	22.70%	71.60%	5.60%	6.10%	Multi/single-family res./commercial
Lemon Grove	Massachusetts Ave.	0%	82.10%	3.50%	0%	Single-family res./commercial
	Lemon Grove Depot	8.40%	16.20%	60.80%	3.20%	Commercial/retail/light manufacturing—Town Center
La Mesa	Spring St.	11.40%	82%	6.90%	5.10%	Multifamily res.—US Navy Housing/Spring Hill Apts.
	La Mesa Blvd.	7.50%	28.40%	47.30%	5.50%	Multistory mixed-use—City Center
	Grossmont	4.50%	8.60%	78%	2.40%	Commercial/retail/institutional—Grossmont Ctr./Hospital
	Amaya Dr.	23.90%	72.20%	14.60%	3.20%	Three-story/multifamily res.—Villages of La Mesa
El Cajon	El Cajon Transit Center	10.40%	26.90%	6.50%	6.40%	Industrial
	Arnele Ave.	5.80%	21.30%	26.10%	0%	Retail/auto dealership/light industrial—Parkway Plaza
	Weld Blvd.	0%	0%	1.50%	0%	Undeveloped/industrial/airport—Gillespie Field (Airport)
Santee	Santee Town Center	0.50%	10.80%	23%	0%	Undeveloped/power retail

[a]Zoning categories are, from left to right, percentage of land within one quarter mile of the station zoned multifamily residential, any residential, commercial, and mixed use.

Table 8.2: Transit-Oriented Projects in San Diego County (transit-based residential)

Project	Station	City	Year Completed	Project Size (# units)	Project Description	Redevelopment Area	Developer	Size of Development	DU/Acre or FAR
Creekside Villas	47th St.	San Diego	1989	141	Two-storey apartments	Southeastern	WKB General Partnership	4 acres	35.3 DU/acre
Barrio Logan	Barrio Logan	San Diego	1989	144	Two-storey townhouses	Barrio Logan	MAAC[a]	4 acres	36 DU/acre
Villages of La Mesa	Amaya	La Mesa	1989	384	Two- and three-storey apartments	Alvarado Creek	Douglas Allred Co.	19 acres	20 DU/acre
Navy housing	Spring St.	La Mesa	1989	244	Two-storey apartments	Eastridge	US. Dept. of Defense	38.5 acres	6.34 DU/acre
La Mesa Village Plaza	La Mesa Blvd.	La Mesa	1991	95	Four-storey; mixed use: condos and retail/office	Downtown	Common-wealth Co's.	5.4 acres	17 DU/acre
Rio Vista West	Friars Rd.	San Diego	Under construction	679–1070	Mixed use: residential/retail/office	Private ownership	Calmat Properties Co.	94.5 acres	25–39 DU/acre[b]

DU = dwelling unit; FAR = floor area ratio.
[a]MAAC is the nonprofit Metropolitan Action Advisory Committee, which sponsors various projects in low- to moderate-income communities.
[b] For residential component, based on the Rio Vista West Specific Plan.

consistent with TOD. First, we considered all projects that had been identified, either by the literature or by San Diego County planning directors, as having TOD elements. We also included projects that were mentioned in promotional literature from the MTDB (Metropolitan Transit Development Board, n.d.-a and n.d.-b) and local specific plans (e.g., City of San Diego, 1994). From among those projects, we then only selected those meeting the following criteria:

1. The project had to include residential or office/commercial development.
2. The development had to be within a quarter mile of either an existing trolley station, or a station that was under construction as of summer of 1995.
3. Construction of the nearby station or trolley line must have preceded the development of the project. Specifically, we excluded projects that were constructed before planning for the nearby station began.
4. The project must either exist or be under construction by summer of 1995. We excluded proposed developments because of our concern with implementation, as opposed to design or concept development.
5. The project must in some way reflect the influence of the nearby rail station.

In practice, we excluded no projects based on the last condition, since all projects that met the four other conditions reflected at least some attempt to take advantage of the nearby rail station.

The projects in San Diego County meeting the above criteria are listed in tables 8.2 and 8.3, which illustrate two themes that are key to our analysis. First, if one restricts attention to projects that either exist or are being constructed and were built after the rail transit station, there are relatively few TODs near San Diego Trolley stations. To date, the San Diego County experience is consistent with what has happened elsewhere; TOD projects are built in some places but they are not a major trend. Second, the TOD activity in San Diego County is concentrated in two cities—La Mesa and San Diego.[9] Even more notably, all four of the stations in La Mesa have nearby TOD projects. One important question is why TOD development has proceeded farther in La Mesa than in any other city in the County, and conversely, why TOD has had a much smaller impact in cities other than La Mesa and San Diego. We address these issues first by summarizing the implementation of TOD projects in both La Mesa and San Diego, and then by drawing on our interviews with planning directors to analyze why TOD has not made more inroads in other cities.

Transit-Oriented Development in the City of La Mesa

During the 1980s, TOD projects were built near each of La Mesa's four East Line trolley stations. Yet an analysis of the planning process for each

Table 8.3: Transit-Oriented Projects in San Diego County (transit-based commercial)

Project	Station	City	Year Completed	Project Size (square feet)	Project Description	Redevelopment Area	Developer	Size of Development	DU/Acre or FAR
Grossmont Trolley Center	Grossmont	La Mesa	1989	113,278	10 storey, auto-oriented commercial	Fletcher Parkway	CCRT Properties	15.5 acres	Approximately 20% coverage[a]
Mills Building	Imperial and 12th	San Diego	1989	180,000	10 storey; gov. offices, ground-floor retail	Centre City	SD Regional Building Authority	1.38 acres	FAR = 3.0[b]
America Plaza	America Plaza	San Diego	1992	931,510	34-storey; offices, ground-floor retail	Centre City	Shimizu Land Corp.	2.62 acres	FAR = 8.17[b]

[a]Lot coverage for Grossmont Trolley Center is the percentage of the property covered by single-storey development.
[b]Floor area ratio (FAR) is calculated by dividing the gross square footage of the building by the total square footage of the parcel.

TOD project shows that those developments were often pursued not only for their link to rail transit, but also because they supported other local goals.

La Mesa Village Plaza at the La Mesa Boulevard Trolley Station

The La Mesa Village Plaza is a mixed-use project adjacent to the station at La Mesa Boulevard in the city's downtown (figure 8.2). Redevelopment of the property started in the early 1970s, before the plans for the East Line were announced. In the mid-1970s, the redevelopment authority cleared the land and was prepared to sell the property to private interests who proposed building an office complex on the site. When the private developers lost their financial backing, the redevelopment authority was forced to look for other interested parties. The situation stayed that way, with the land cleared and awaiting redevelopment, for several years (Wear, 1997).

In the interim, planning for the East Line extension had begun. During the 1980s, the redevelopment authority issued several requests for proposals, and eventually settled on a mixed-use residential and commercial plan submitted by Commonwealth Companies (Wear, 1997). The planned extension of light rail service had influenced the development possibilities at the site, but there were also other factors at work. As Dave Wear, planning director at the time (and now La Mesa City Manager), recalls, the empty parcel in the city's downtown was "embarrassing." There was

Figure 8.2. Photo from La Mesa Blvd. (Downtown) Station.

an imperative to do something with the parcel, and the redevelopment agency viewed the Commonwealth proposal as the highest and best use for the property (Wear, 1997). The development, which opened in 1991, features 95 condominiums, offices, and ground-floor retail.

The Grossmont Trolley Center at the Grossmont Station

The Grossmont station, like the station a half mile away at Amaya Drive, is within the Grossmont Specific Plan area. The specific plan was developed in the mid-1980s to take advantage of the intersection of the 8 and 125 Freeways, the then proposed light rail stations at Grossmont and at Amaya Drive, the Grossmont Shopping Center regional mall, and Grossmont Hospital, at the time the city's largest employer (City of La Mesa, 1985a).

Topographic constraints greatly influenced the development opportunities along the portion of the East Line within the Grossmont Specific Plan area. The line follows a creek bed in a valley approximately forty to fifty feet below the existing development. This elevation differential limited the ability to link the Grossmont station to either the nearby Grossmont Shopping Center regional mall or the Grossmont Hospital (City of La Mesa, 1985a; Wear, 1997). The difference in elevations also created a buffer between the stations and preexisting land uses. Partly for that reason, there were very few NIMBY ("not in my backyard" protest) issues involved in either the Grossmont Trolley Center Development at the Grossmont station or the Villages of La Mesa development at the Amaya Station. As Dave Wear said, "We've had potentially controversial development proposals go through here [the Grossmont Specific Plan area] without a peep" (Wear, 1997).

Another important consideration was that the flood plain along a portion of the creek bed had been a local development priority since before construction of the East Line began. In the case of the Grossmont Center project (adjacent to the Grossmont station), the specialty retail built there allowed the City of La Mesa to use the vacant land along the creek bed, which facilitated long-standing local development goals for that property (Witt, 1995).

Villages of La Mesa at the Amaya Station

In the mid-1980s, the La Mesa redevelopment authority owned vacant property next to the site of the proposed Amaya station (figure 8.3). The land for the Amaya station had been purchased by the MTDB some years earlier. Both the redevelopment authority's property and the MTDB's land were split across different elevations, with some of each property on the valley floor and some of both parcels at a higher elevation to the north. The City of La Mesa proposed to the MTDB that they swap part of their respective properties, so that both the MTDB's station and the redevelopment authority's proposed apartment complex could be built on flatter parcels that would require less grading (City of La Mesa, 1985b;

Figure 8.3. Photo from Amaya Station.

Wear, 1997). The city suggested that the swap would lower construction costs for both parties. The land swap was completed in the late 1980s, and the 384-unit Villages of La Mesa apartment complex opened in 1989.

Importantly, the redevelopment authority's property near the Amaya station had become an unofficial dump site over the years. La Mesa's motivation for developing the Amaya site predates the light rail line. As both Dave Wear and Dave Witt noted, the Amaya property had been zoned multifamily residential since before the East Line was proposed. In the eyes of the city, the opportunity to develop the property and remove the eyesore of a vacant dump site was welcomed, irrespective of links to the rail transit system (Witt, 1995; Wear, 1997).

Navy Housing at the Spring Street Station

Trolley and freeway access were both important for the Spring Street station development (Witt, 1997). The Navy housing that opened there in 1989 was developed by the Department of Defense. The city did not provide much planning leadership in this project. The property is on a steep, granite slope, which increased development costs. Dave Witt, the current planning director in La Mesa, has suggested that "the Defense Department was the only [entity] that could afford to build on that property" (Witt, 1997). The 244-unit complex opened in 1989.

Summary of La Mesa's Transit-Oriented Development

Our interview results revealed that most of the goals of TOD projects in La Mesa went beyond transportation goals. Specifically, the City wanted to develop the long vacant downtown parcel at La Mesa Boulevard, use the property in the flood plain near the Grossmont station, and remove the vacant dump site near the Amaya station. Witt stated, "I think that may have been partly why some of these projects [in La Mesa] are on the ground, because they weren't being driven, even for the most part, by the Trolley" (Witt, 1995). Thus, one reason why La Mesa has more TOD projects than many other San Diego County cities is that TOD helped facilitate some of La Mesa's long-standing goals.

Yet the convergence of local goals and TOD in La Mesa, while important, might not be the only element in the development of that city's transit-oriented projects. La Mesa has had continuity among top planning officials for several years. Dave Wear was Director of Community Development in the City from 1980 until 1990. In 1990, Wear became City Manager, Dave Witt (previously the Assistant Director of Community Development) became Director of Community Development, and both remain in those positions to this day. This appears to have provided some stability and continuity that possibly facilitated some of the development projects in La Mesa. Also, La Mesa officials seemed to approach each project with a practical orientation. Director Dave Witt viewed many of the projects as "just good planning" (Witt, 1995).

If anything, the vision of TOD in La Mesa was informed more by what was consistent with local goals than by a comprehensive view of TOD. This possibly made city officials more willing to work out the details required to facilitate developments (such as the land swap at the Amaya station) and might have provided local political support for the TOD efforts.

Transit-Oriented Development in the City of San Diego

The City of San Diego has two transit-based residential projects (Creekside Villas and Barrio Logan), two major downtown transit-oriented office buildings (the Mills Building and America Plaza), and the initial commercial portion of a proposed mixed-use TOD at Rio Vista West. This TOD activity reflects the fact that San Diego is by far the largest city in the county, with the bulk of the trolley stations. The TOD developments to date also reflect a commitment to transit-based land-use planning on the part of the City of San Diego, the MTDB, the San Diego Redevelopment Authority, and the Centre City Development Corporation, the redevelopment authority for much of downtown San Diego.

The Creekside Villas apartments, which opened in 1989, provide mostly low-income housing. The project was developed by WKB General Partnership on land leased from the MTDB. The development includes a daycare center that can accommodate forty-four children. This reflects, in part, a commitment on the part of both the City of San Diego and the MTDB to provide important work-related destinations, such as daycare, near transit-based residential and office projects (Metropolitan Transit Development Board, n.d.-b).

The Barrio Logan Redevelopment Area was established in 1991. The Barrio Logan neighborhood is a lower income, predominantly Hispanic community south of downtown. In 1992, the Mercado Apartments opened within a quarter mile of the Barrio Logan station. The 144 apartments were targeted for low- and moderate-income families (Metropolitan Transit Development Board, n.d.-a). To finance the Barrio Logan residential project, the redevelopment authority pulled together a public-private partnership that includes six entities. Notably, the San Diego Redevelopment Agency provided a land write-down and subsidies that were worth close to $2,000,000. Affordable housing program tax credits accounted for just over $5,000,000 (Bernick and Cervero, 1997, pp. 260–262).

In downtown, both the Mills Building and America Plaza have trolley stations built into the structure. The Mills Building is a 10-storey project that houses the headquarters of the MTDB. This building, at the Imperial and 12th station, was built by a joint powers agency, composed of the County of San Diego and the MTDB, in partnership with Starboard Development Corporation (Metropolitan Transit Development Board, n.d.-b). The MTDB played a pivotal role in this development, both by providing the land and by siting their headquarters in the building. The transfer station for the East and South Lines opens onto the building's lobby, providing a close link between the government offices, trolley transfer nodes, and ground floor retail, which are all part of the project.

America Plaza is within the Centre City Redevelopment Area, yet the project required no direct public assistance (Bernick and Cervero, 1997). The America Plaza Transfer Station is part of the ground floor of the building. The thirty-four-storey office tower (the city's tallest) was developed by the Shimizu Land Corporation in the late 1980s and opened in 1992 (Metropolitan Transit Development Board, n.d.-b). In addition to office space, the project includes a retail galleria and food court and plans for a future hotel (Metropolitan Transit Development Board, n.d.-b).[10] Like the Mills Building, America Plaza is on the Centre City portion of the trolley system, which forms a loop around downtown San Diego.

The Rio Vista West project is on the Mission Valley Line, which opened in November 1997. The entire 95 acres is owned by CalMat properties (City of San Diego, 1993). The Rio Vista West development, designed by Peter Calthorpe, features a mix of residential units and commercial properties, and incorporates many pedestrian-friendly design elements (City of San Diego, 1993). The downturn in the southern California economy during the early 1990s slowed the project's implementation, and by 1995 the developer chose to build an initial phase of auto-oriented, "big-box" retail on the site to provide income for future development.

Overall, while these developments represent more projects than exist in La Mesa, they are the outcome of a process reflecting both San Diego's commitment to TOD and some of the difficulties in implementing transit-oriented projects. The City of San Diego has arguably devoted more planning resources toward TOD than has any other city in the county, including La Mesa. Yet the planning director in San Diego also suggested that the city faces many of the same barriers cited by the cities in San Diego County without TOD (Freeman, 1995). In the City of San Diego, the barriers to TOD are most evident along the South Line, which passes through lower income residential and industrial neighborhoods and where, except for the Barrio Logan project, there has been no redevelopment oriented toward rail transit. More generally, the experience in the City of San Diego illustrates the importance of understanding not only the local commitment to the TOD idea, but also the barriers that often stand in the way of implementing that idea.

Barriers to TOD Implementation in San Diego County

As described in chapters 6 and 7, the literature has considered several possible barriers to transit-oriented development. A brief summary of possible TOD barriers, consolidated from the ideas mentioned in chapters 6 and 7, Deakin and Chang (1992), and Cervero, Bernick, and Gilbert (1994), is listed below:

1. Existing land-use patterns near rail stations can constrain the opportunities for TOD.
2. Difficulties in assembling large parcels of land can limit TOD opportunities.
3. The private land market might be unable to sustain new development projects, including transit-oriented ones.
4. Local economic and fiscal impacts might discourage localities from pursuing TOD.
5. Local officials might not be adequately educated in both the regional advantages and local impacts of TOD.[11]

Below we summarize the findings from our interviews of all planning directors in relation to the role of each of the five possible barriers toward TOD development. For a more detailed presentation of the interview results, see Boarnet and Compin (1996) and Compin (1996).

Constraints Imposed by Using Existing Right-of-Way

All stations currently served by the San Diego Trolley, except for the Santee station and the seven stations on the Mission Valley Line, were sited along existing rights-of-way. This creates several potential problems for TOD. First, because most trolley stations were sited in areas with existing development, the scope for adding new transit-oriented projects is limited by the ability of the city or other entities to redevelop the area. The planning directors in Chula Vista, El Cajon, Lemon Grove, and National City noted that the land near their Trolley stations is already developed, and they do not expect any substantial land-use change near those stations.

Second, the preexisting development near existing stations is often auto oriented rather than transit oriented. For example, Robert Leiter, planning director for Chula Vista, noted that the South Line parallels the Interstate 5 corridor in his city (figure 8.4). Land uses along that corridor, in Leiter's view, were influenced much more by the preexisting freeway than the rail transit line. Mr. Leiter stated, "I would say probably if you went back in history, I would think that I-5 and even before that, Broadway—the previous main North/South arterial that went through Chula Vista—had a lot more to do with determining the land use patterns in that area than the transit stations per se" (Leiter, 1995).

The third difficulty with using existing right-of-way is that, in some cases, the character of land uses near the stations was determined by preexisting freight rail, and is not conducive to residential development (figure 8.5). Roger Post, of National City, speculated that the manufacturing and commercial uses near his city's South Line stations could be "a major negative" for any plan that proposed to add residential development nearby (Post, 1995).

TOD Implementation and the Availability of Undeveloped Land

Given that many San Diego Trolley stations are in already-developed areas, land assembly is an important issue for TOD projects in San Diego County. As table 8.2 illustrates, all of the TOD projects in San Diego except Rio Vista West are within redevelopment zones. Redevelopment zones, as authorized by California's Community Redevelopment Act, are both land assembly and tax-increment financing tools.[12] The exception to the use of redevelopment zones, Rio Vista West, also illustrates the

Figure 8.4. Photo of Bayfront E St. Station in Chula Vista.

Figure 8.5. Photo of Harborside Station/National City Shipbuilding.

importance of land assembly. That project is being constructed on a 95-acre parcel that is owned by CalMat Properties, and historically has been a sand and gravel operation (City of San Diego, 1993). That property had been vacant before the planning for Rio Vista West began. Three other TOD projects in the county (Villages of La Mesa, Spring Street Navy housing, and the Grossmont Trolley Center) have been built on previously undeveloped parcels. This suggests the importance of undeveloped land when building TODs, consistent with the interview results that were summarized in the preceding subsection.

The Role of Market Forces

Development of all kinds slowed considerably in California during the recession of the early 1990s. All of the planners, except Dave Witt of La Mesa, referred to the slow economy when discussing the prospects for high-density uses near their stations. The general conclusion was that the economic downturn, and decreases in land and property values in the region, limited prospects for intense development. Even in La Mesa, which built many TOD projects before the downturn, the recession had an effect. The original Grossmont Specific Plan (City of La Mesa, 1985a), written in 1985, envisioned high-density uses that, according to both Dave Witt (1997) and Dave Wear (1997), could not be supported in the 1990s land market. The revised Grossmont Specific Plan (City of La Mesa, 1994) was much more modest in terms of the proposed increase in both commercial and residential densities in the area.

Fiscal Impacts of TOD

Chapter 7 suggests that residential development near rail transit stations might bring adverse fiscal impacts for localities. To recap, there are two main reasons for this. First, in California, a portion of the sales tax revenues that are generated within each city are returned to the cities, such that land uses that create taxable transactions (i.e., commercial) are attractive from a fiscal perspective. Second, many cities perceive that medium- and high-density residential developments create service and spending obligations that exceed the tax revenue generated from those projects. These fiscal pressures are reflected in statements made by all of the planning directors except Dave Witt of La Mesa.

For example, James Griffin of El Cajon noted that "commercial has the advantage certainly of generating sales tax" (Griffin, 1995). Similarly, Niall Fritz of Santee said, "We need to get the highest and best return in order to continue to provide other services to the people who live here. And that means today; not tomorrow. So today we're going for retail uses. We do not have the luxury to wait for tomorrow" (Fritz, 1995).

Education About Regional TOD Goals

Our interviews suggest that the education of planning directors about TOD is not a problem in San Diego County. Both the San Diego MTDB and San Diego Association of Governments (SANDAG) have promoted TOD in the region. The MTDB in particular has put together a strong TOD public relations campaign, including a film, brochures, and informational meetings with local government officials and general plan advisory committees. All of the planning directors interviewed showed in-depth knowledge of even the finer points relating to TOD. They used TOD terms freely and knowledgeably and were familiar with the theoretical basis for TOD.[13]

While all planning directors stated that they agreed with the regional goals for rail transit put forth by MTDB and SANDAG, *each also made it clear that local goals came first with respect to land use.* Our interviews further suggest that the barriers listed above are often (but not always) impediments to TOD implementation, and that education, by itself, will not overcome structural factors such as preexisting development, land availability, market forces, and fiscal pressures.

Summary of the Barriers to TOD

In the five cities with no existing TOD (Chula Vista, El Cajon, Lemon Grove, National City, and Santee), there have been several barriers to TOD. Most important, the rail lines in those cities do not pass through properties that have been slated for redevelopment. Instead, they often pass through areas already oriented toward the automobile (along the portion of the South Line that parallels Interstate 5) or places where existing development was incompatible with residential projects.

Still, our interviews suggest that no single barrier is overwhelmingly important in explaining the limits of TOD outside of La Mesa and San Diego. The constraints imposed by existing rail rights-of-way, land availability, market forces, and fiscal pressures all seem to play a role. Overall, no locality appeared hostile toward rail transit or regional rail goals. Instead, each city had local goals that took precedence. This was true even in La Mesa, but as mentioned above, local goals in La Mesa were more compatible with TOD than in other parts of the county.

Summary

The TOD idea is a comprehensive attempt to use land-use planning to enhance the viability of rail transit. As with many new ideas, most authors have focused on the broad vision, either by enumerating design guidelines (e.g., Calthorpe, 1993), evaluating the transportation impacts (e.g., Cervero, 1994a), or describing how TOD might form the basis for a new approach to rail transit planning (e.g., Bernick and Cervero, 1997). With so much focus on the "big picture," it is easy to assume that TOD is a "big idea" that will be implemented according to a comprehensive strategy.

Yet the lesson from San Diego County is that progress toward TOD goals is often made in incremental steps. TOD projects are the outcome of a number of local governments acting in their own interests, pursuing opportunities as they present themselves and working within local constraints. Where local conditions are consistent with TOD, as was the case in La Mesa, progress can be somewhat rapid. Elsewhere, barriers and competing local concerns carry the day. This is consistent with the incremental process of policy implementation outlined by Lindblom (1959). While for any station or city, each project is a significant effort that requires much strategic planning, the character of station-proximate land use throughout the San Diego Trolley system is adapting slowly. This is also consistent with the conclusions of Knight and Trygg (1977), who found that development occurred near rail stations when other factors, such as strong market demand, low-cost available land at attractive sites, and supportive land-use policies, are in place.

Nothing indicates that the experience of San Diego County has been unusual. The barriers to TOD have been cited before in the literature, and the experience with TOD in San Diego is in many ways consistent with that documented elsewhere (e.g., Bernick, 1990; Cervero, Bernick, and Gilbert, 1994). As mentioned above, if anything San Diego might be expected to be more conducive to TOD than many urban areas, so the barriers identified in San Diego County might be even more important elsewhere.

The results of this case study suggest that those barriers to more extensive TOD implementation in San Diego County are fourfold—the constraints imposed by existing rights-of-way, difficulties assembling land in already-developed areas, market conditions, and fiscal and economic incentives. Of those, the condition of the land market and fiscal incentives are largely beyond the reach of local governments.[14] The primary way for regional and local governments to facilitate TOD would thus appear to be aligning rail transit systems in ways that make TOD more feasible. One lesson from this research is that the legacy of preexisting land uses is an important determinant of TOD implementation, and thus TOD prospects are heavily influenced by the alignment of a rail line and the placement of stations.

San Diego will, in the future, provide an opportunity to study the link between system alignment and TOD. The Mission Valley Line, under construction at the time these data were collected, is the first trolley line not using preexisting freight right-of-way. Instead, the MTDB chose to put the Mission Valley Line through a high-growth corridor to the north of downtown. Mission Valley contains regional shopping malls, major hotels, office and commercial complexes, a major league sports stadium, and new residential development (City of San Diego, 1994). This preexisting development, plus the fact that the corridor is still growing and has developable land, makes the Mission Valley Line a more attractive place for development projects than much of the South and East Lines. Many of the future TOD plans in the City of San Diego are on the Mission Valley Line, although only one of those TODs (Rio Vista West) was being built at the time of this study (City of San Diego, 1994). By siting the line along an existing growth corridor, the MTDB has possibly created a situation where TOD is consistent with other local development plans, much as it was in La Mesa in the mid-1980s.

This is not necessarily an optimistic lesson for TOD proponents. Placing rail lines along high-growth corridors can be an expensive option, especially when those corridors do not have suitable existing rail rights-of-way. Thus, the benefits of TOD might be best enhanced by expensive rail alignment decisions. Whether the incremental gain in, for example, transit ridership outweighs the cost of placing a line along a high-growth corridor remains an open question. Local and regional authorities should be aware of this trade-off, and should carefully evaluate the costs and benefits of particular rail alignments.

More generally, an unambiguous conclusion from this study is that TOD implementation proceeds more smoothly when it is consistent with other local development goals. Regional rail officials should be aware of that, and should use that information both in assessing the prospects for TOD and in deciding which proposals are the most promising development opportunities.

More generally, the San Diego experience suggests that the transit-oriented implementation process is, by its nature, a slow one. If transit-oriented planning can enhance the prospects of rail transit, officials should realize that any benefits might take years to be realized. One lesson from San Diego County is that, over fifteen years after light rail service began, the number of existing TOD projects is small and concentrated either in a city where TOD was consistent with other goals or in the region's largest city. To count on near-term systemwide ridership impacts from this one planning strategy would be risky, both because of the questions about the link between ridership and TOD discussed in chapter 7 and because of the incremental implementation process we documented in San Diego County.

The experience in San Diego County illustrates the importance of focusing not only on the TOD vision, but also on the details of how such plans are actually implemented. Regional authorities should attempt to understand factors such as market demand, land availability, fiscal pressures, and local goals that have influenced TOD in San Diego County. Furthermore, local and regional officials should pursue careful assessments of demonstrated TOD benefits and project costs. The lesson for regional authorities is that localities might already be doing that, at least in broad terms, and that these projects are pursued most aggressively when consistent with local goals. This implies a process likely to look like the slow, incremental implementation in San Diego County.

PART IV
What Role for Travel by Design?

We began this book with three broad questions about the tendency to link urban design and transportation planning. First, will the new designs yield the expected transportation benefits? Second, can the designs be implemented on a scale large enough to meet their transportation policy goals? And third, are these proposals good ideas? In this chapter we summarize our results and consider their implications for research and policy.

9 Lessons for Research and Practice

Transportation problems seem to offer no end of interesting policy wrinkles and technical challenges, but despite the promise of each new technological innovation, financial windfall, and dazzling social science breakthrough, planners have not fared well. Air pollution, fuel, and traffic congestion costs continue to mount to where the benefits of making any headway appear substantial. Yet as more freeway lanes are dedicated to car-poolers and tollways, and new transit systems continue to soak up many billions of dollars, getting people to "improve" their driving behavior remains the ultimate planning brick wall. Increasing evidence suggests that transportation demand management schemes have extremely limited effectiveness, in the sense that only marginal and perhaps even cost-ineffective changes can be expected from most of the tools applied thus far.

One view is that the planner's arsenal of transportation demand management tools has proven largely ineffective in dealing with traffic congestion especially. The somewhat more optimistic account of some planners and architects is that attention has been focused on symptoms rather than the disease itself. As discussed in chapter 1, the vanguard of such urban design schools as the New Urbanism, Neotraditional planning, and transit-oriented development collectively argue that the way we organize space has profound implications not only for traffic patterns but perhaps also for our sense of self and modern civilization as a whole. Prominent urban designers, planners, and political leaders forcefully claim that these development strategies will, among other things, improve traffic conditions, reduce home prices, and generally increase the quality of residential life.

Of course, this is just talk. As bold and stirring as these claims may be, they are mainly meant to get us thinking afresh about where and how improvements can be made—not as cold hard facts. Most transportation planners probably recognize that blanket statements of this nature are overly simplistic. Even the architects and planners promoting these ideas are usually careful to emphasize the many ingredients necessary to obtain desired results: the straightening of streets to open the local network, the calming of traffic, the better integration of land uses and densities, and so on. The new designs have many elements, and while their purported transportation benefits are often featured, they are by no means the only component.

Nonetheless, these ideas have had a great impact on modern city planning thought and practice. A growing number of general and specific plans feature various combinations of these elements as self-evident improvements, and the claim that virtually any one such element has transportation benefits has rarely been challenged in either the practitioner or scholarly literatures.

In this book we asked, *Can these strategies work, can and will they be implemented,* and *are they a good idea?*

The Influence of the Built Environment on Travel

The appeal of the new designs is hard to deny, but can they deliver? We seem not to know. Surprisingly, there is little credible knowledge about how urban form influences travel patterns. Given the enormous support for using land use and urban design to address traffic problems, it was somewhat surprising in chapter 3 to find the empirical support for these transportation benefits to be inconclusive and their behavioral foundations obscure. Prior evidence on the link between design and travel is difficult to interpret and tells us relatively little about the behavioral nature of the problem and thus provides a weak foundation for policy advice.

Our contribution involved analyzing travel within a standard demand framework, in chapter 4, and then testing how various land use and urban design features influence trip costs and ultimately travel behavior, in chapter 5. Little is controversial in our approach, yet it has been surprisingly absent from recent discussions of this topic.

Our results indicate that under fairly general conditions, a mix of changes in urban design either increase, decrease, or do not change car trip frequencies and vehicle miles traveled (VMT). The only design feature that appears to unambiguously reduce car trips and thus VMT, though by how much we cannot say, is traffic calming. While VMT is unlikely to rise, or rise much, that unexpected things are plausible should provide considerable caution for persons who wish to use other design elements to influence driving behavior.

The point is not that most anticar plans can unintentionally backfire, though they might, but that the link between nonwork automobile travel and urban design is complex—in particular it is not as simple, obvious, or deterministic as much popular urban design rhetoric implies. This is not to say that there is no link, or that the link works substantially differently from the hypotheses of advocates of the new urban designs. In fact, some of our evidence in chapter 5 supports the argument that land use does influence nonwork travel. Yet those tentative links are neither self-evident nor based on a priori theory. Instead, the pattern is complicated and will vary depending on the particulars of various urban areas. Overall, the new urban designs, rather than being an organizing framework for transportation planning, are a set of hypotheses that should be seriously examined within a coherent research framework. Rather than *facts* regarding how design will influence travel, they are researchable *questions.*

That said, the empirical results from chapter 5 do suggest a few rules of thumb that can help guide policy. First, planners should carefully focus on how urban designs influence nonwork car trip speeds and distances. Intuitively, urban designs that place nonwork destinations closer to residences might both reduce nonwork trip distances and, if streets congest or if planners specifically try to "calm" car traffic, slow automobile trip speeds. These two changes—slower automobile trip speeds and shorter trip distances—can have opposing effects on trip generation rates. The evidence from the San Diego data in chapter 5 suggests that automobile trip generation will be reduced in instances where automobile trip speeds are slowed to the point that they offset the tendency of shorter trip distances to induce more car trips.

An implication for planners is that the built environment, if it is to reduce nonwork car trip making, should slow automobile speeds sufficiently to counteract the effect of any reduction in trip distances. Oddly, then, congestion can be an ally of planners who seek to eliminate automobile trips, as some reduction in trip speeds, possibly due to increases in congestion created by higher densities, can provide an incentive for persons to avoid driving.

This is less unusual than it might seem, as many policies seeking to reduce car travel, including carpool lanes and mass transit, rely to some extent on congestion levels to reduce the attraction of either driving or (in the case of carpool lanes) driving alone. The difficulty is that congestion itself is an external cost and policy-makers must balance the extent to which congestion aids plans to reduce driving with the benefits of reducing traffic congestion. The proper way to evaluate that balance is within a benefit-cost framework. More generally, urban plans in already-congested areas, such as central cities or more dense inner suburbs, might be more likely to deliver on their transportation promises than the same designs in outer suburban areas.

A second rule of thumb suggested by our research is that shortening trip distances sufficiently to facilitate walking is problematic. The evidence suggests that most persons are willing to walk not much more than a quarter mile for most trips (Untermann, 1984). Thus, planners would be well advised to remember that until trip distances are shortened to this distance or less, other design strategies might do as much to encourage driving as travel by other means, walking included.

The last rule of thumb is in some sense the murkiest, as it does not flow directly from the research in chapters 4 and 5. Intuitively, the pattern of trip chaining that includes how one combines trips for different purposes, often as part of a commute, has much to say about the link between urban design and travel behavior. Some new urban designs seek to cluster trip destinations, and while our analysis has focused exclusively on clustering around residential locations, the same concentrations could be encouraged near job centers. A largely unanswered but important question is how different urban designs do or do not influence trip-chaining patterns, and in turn how that informs transportation policy.

Neighborhood Supply Obstacles

The question of the supply of less car-dependent neighborhoods involves two issues—the market potential for these designs, and whether local regulatory regimes allow that market to meet whatever demand there may be. Market potential is a serious and yet unresolved issue. Chapters 4 and 5 demonstrated that we do not know how many car trips (if any) these developments might take off the road. Hence, unless the magnitude of the travel behavior effect is dramatically large, the new urban designs will have to be implemented on a reasonably broad scale to be a major component of air quality or congestion management plans.

The evidence, thin as it is, suggests that the scope for the new urban designs is more limited. Certainly some persons might wish to live in moderately dense, Neotraditional, pedestrian-oriented villages, but there is little that suggests that those developments will soon become more than a niche market. Yet debating the magnitude of the market share leads us to ask the wrong questions. Rather than agonizing over how many households might choose to live in New Urbanist or similar neighborhoods, policy-makers should examine whether the land market supplies those neighborhoods to persons who want them. Measuring the transportation benefits of those neighborhoods can follow later.

Putting this another way, a nascent movement in planning has begun to ask whether land-use regulations, especially zoning, interfere with the desire of people to drive less. We recast that question in chapters 6 and 7 by asking whether local governments have incentives to not supply certain kinds of neighborhoods for the "wrong" reasons. If so, how might such undersupplies be addressed? This approach moves the focus from such tasks as encouraging persons to buy housing in dense neighborhoods—a political nonstarter and a problematic issue generally—to the question of whether or not the housing market is constrained by regulatory policy from adequately providing more dense neighborhoods.

Our examination of the incentives of local governments used the example of transit-oriented development (TOD) in southern California. Studying TOD had several advantages, but three are most important. First, the potential neighborhood site is, by definition, adjacent to a rail station. This provides a clear locational focus for our analysis. Second, while Neotraditional and New Urbanist neighborhoods are still few and far between, there have been several attempts, even in southern California, to build transit-oriented neighborhoods, providing a growing data base for empirical and case analysis. Third, given that the land near rail transit stations often includes potential sites for commercial and office developments, the economic and fiscal issues that influence local incentives toward land use appear to be especially stark in the case of TOD. While examining the narrow realm of municipal land-use behavior in relation to TOD, we can gain insight into broader issues of whether and how local governments might constrain the supply of some kinds of neighborhoods that might otherwise meet market needs.

We found that localities respond to both economic and fiscal incentives in their plans.[1] Both zoning maps and the interview responses from planners indicate that local governments view transit stations as opportunities for office and commercial development, and will favor those developments at the expense of other possible land uses. Part of the motivation for this is fiscal and is tied to the way that land uses create revenues for local governments. Broader economic concerns also play a role.

One conclusion is that, in addition to thinking about market demand, advocates for transit-oriented land-use planning must consider how well those plans match the incentives of municipalities. Another is that myopic regulatory barriers to supply should be removed or diminished. Again, however, it will be necessary to understand municipal behavior, and possibly change municipal development incentives, before progress will be possible.

For example, Tiebout (1956) conjectured over four decades ago that municipalities within an urban area would specialize to attract residents. Each locality would attempt to carve out its own market niche. Under certain extreme assumptions, this cross-municipality competition would meet the varied demands of consumers for different types of neighborhoods with different levels of local public services. One way to recast the supply question, then, is why Tiebout competition might not lead to an efficient supply of desired neighborhood types, and whether some kinds of neighborhoods tend to be undersupplied in the outcome. Future research and policy would benefit from more attention to this issue.

Is Urban Design Good Transportation Policy?

The transportation objectives of urban design are most appropriately analyzed within a traditional regulatory policy context. Along those lines, the theory of externalities provides a useful framework. The problem with cars is not that they are bad as such, but that car travel brings with it undesirable side effects *for which the market does not provide compensation.* These externalities include air quality problems, traffic congestion, and undesirable impacts on neighborhood quality of life.

An implicit argument for using urban design as transportation policy is that the externalities associated with automobile travel are pressing problems, other policy alternatives have been exhausted, and these design strategies will achieve the desired result. The discussion in chapter 2 cast doubt on the first two points, especially regarding air quality and traffic congestion. Environmental regulations have produced cleaner air in most urban areas. While measures of traffic congestion based on road capacity indicate worsening problems in many metropolitan areas, other data indicate that commute times have hardly changed over the past few decades. Overall, air quality and congestion problems might still constitute pressing policy issues, but the often unstated assumption that these problems are rapidly growing worse appears wrong.

Of course, new environmental problems remain on the horizon and might change this optimistic assessment (e.g., Getter, 1999). Based on a recent study of the Los Angeles air basin, the cancer risk from air pollution is much higher than experts previously suspected. The contribution of automobiles to the emission of greenhouse gasses has only recently received attention, and future policy activity in that area should depend crucially on assessments of the link between climate change and carbon emissions, and (ultimately) on assessments of the cost of the external harm from carbon emissions.

We do not mean to minimize either issue here, as both could potentially prove to be as important, or even more important, for the next several decades of policy as more traditional air quality and congestion problems were in past years. Still, our analysis of the transportation benefits of particular urban design features has, like most other analyses before, focused on more traditional transportation issues. This is largely because those are well understood and because reasonable estimates of the value of pollution and congestion costs exist. Furthermore, in many instances, policy activity for the "newer" environmental problems will likely be evaluated against the alternative policy options that already exist, and so we argue that it is meaningful and important to assess how urban design compares to other existing regulatory options.

First note that land-use and urban design strategies lack the flexibility of many other regulatory options. They propose to change the price of travel but they do so by literally building or rebuilding urban communities. In turn, the built environment is long-lived and difficult to change.

Some might view this as an advantage, hoping that alternatives to automobile travel will become "locked in" by building cities in different ways. Yet the difficulty of changing urban form cuts both ways. Building enough new and different neighborhoods to change the character of a metropolitan area can be a lengthy process; our case study of TOD in San Diego suggests that implementation might be measured on a scale of decades rather than years. Furthermore, once a policy option has been locked into the urban form of a city, that lack of flexibility can be a disadvantage if unforeseen circumstances, new information, or changing preferences create the need for changes in policy. Overall, the uncertainty of most policy environments suggests a preference for flexible policies that can be quickly implemented. Urban form comes up considerably short on that count.

Moreover, many other policies can potentially deliver the transportation benefits sought. For air quality, continued emphasis on engine technology should yield incremental emissions reductions. Taxes on gasoline, driving, or emissions would all encourage persons to drive less. Finding ways to take the most grossly polluting vehicles off the road can potentially provide a large amount of cleanup by focusing on a small number of the vehicles being driven.

For congestion, the most attractive policy is to charge fees to drive on congested roadways. Initial experience with peak period pricing in southern California has set off of wave of proposed experiments across the country, and this augers well for the possibility of controlling urban traffic congestion in the future. While many analysts argue that congestion pricing will remain politically infeasible for decades, if not longer, these policy alternatives are more flexible than urban design, and can typically be implemented in both individual cases and on a broad scale much more quickly.

Neighborhood quality of life, on the other hand, remains a key urban design policy objective. The links among urban design, transportation, and quality of life may be more appropriate than the links drawn to either air quality or traffic congestion (*Consumer Reports,* 1996; Fulton, 1996). While the travel impacts of virtually any urban design are still uncertain, and thus the link between travel and quality of life is unclear, broader aesthetic and public participation goals might indeed be consistent with efforts to improve the sense of place and quality of life within communities. There are also fewer policy alternatives to urban design for this purpose.

Having said that, urban design strategies continue to suffer from our lack of knowledge about their travel impacts, the ability of markets to build alternative neighborhood prototypes, and the willingness of local governments to facilitate their development. Those gaps in our knowledge, coupled with the availability of regulatory alternatives, make urban design as much a research agenda as a policy initiative.

Policy Implications: Incentive versus Outcome Regulation

Neighborhood design standards and transportation infrastructure projects, because they are so long-lived, are inseparable from city building more generally. A city's form, and perhaps its spirit, is shaped in part by its design and transportation infrastructure. Thus, some persons are inclined to confound two somewhat distinct goals—building city forms that will endure and thrive for decades and managing the more quickly changing transportation problems of today and tomorrow.

But we do not know as much as we would like about the travel impacts of one design versus another, including unintended consequences, and other transportation policies might offer better solutions to particular problems. We therefore recommend that planners should lower their expectations regarding the travel benefits of urban design, and perhaps expand their understanding of other transportation policies. Furthermore, we suggest that transportation policy problems are still best attacked singly, because the solutions to problems as diverse as air quality, congestion, and neighborhood quality of life strenuously resist "one size fits all" solutions.

Still, urban design strategies should be included within a suite of policy tools. Among others, pricing the automobile to reflect the full social cost of trips (including air quality and congestion externalities) is an attractive, if politically problematic, policy. Importantly, full-cost pricing of automobiles, by raising the price of car travel, should make less auto-dependent neighborhoods more attractive to consumers. More generally, several policies can play a role in producing cleaner air, less congested roads, and more vibrant neighborhoods. Urban design will likely play a role, but planners should also remember the importance of other policies that can be implemented rather more quickly and adjusted more readily.

Consider the respective roles of incentive regulation and outcome regulation in the control of environmental problems generally, and how they inform the strategies described above. In chapter 2 we noted that the overwhelming majority of environmental regulation in the United States has focused on outcome regulation. Maximum allowable pollutant concentrations, required emissions technologies, and additional highway capacity intended to relieve congestion are all examples of regulations intended to somewhat directly achieve desired outcomes. The use of outcome regulation is understandable in the face of identifiable problems and a pressing need to "do something."

Alternatively, price regulations do not prescribe the desired outcome; they establish a socially desirable set of incentives that accounts for the environmental harm inherent in a specific problem. The use of prices calculated to reflect external (or environmental) harm can make explicit the link between regulatory activity and the crucially important assessment of the cost of a particular type of environmental harm. Prices, as we have discussed, can be adjusted more easily than many other regulatory instruments, and the monitoring costs associated with many incentive (or price) regulatory schemes are low compared with many types of outcome regulation.

After decades of relying primarily on outcome regulation for environmental policy, both government and environmental advocates are increasingly realizing the advantages of incentive regulations, hence the growing popularity of what are called "market-based" environmental policies, such as emissions trading plans. Most important, market-based approaches recognize that to achieve further progress in environmental regulation, it will be increasingly necessary to change behavior. Changing incentives is a powerful way to do so.

The scope for changing behavior is the subject of this book, and these alternatives, incentive and outcome regulation, comprise the regulatory options in transportation policy. Physical design is more outcome than incentive regulation, by emphasizing the physical environment. In thinking about urban design as transportation policy, we should increasingly think about how drivers respond to prices (including the prices implied by changes in urban designs), how governments respond to their own incentives, and how regulators might use knowledge about both types of incentives to foster a better understanding of the effects of and need for new urban designs. A closer focus on the incentives of both drivers and planners, combined with attention to how urban design can work as part of a suite of policy tools, will provide valuable assistance to those working on solutions to transportation problems.

Directions for Future Research

While reserved in our support of specific design proposals for transportation purposes, we feel strongly that this is an important research area. Several research tasks are important if we are to improve our understanding of how travel is connected to the details and broad patterns of the built environment.

First and most important, the link between the built environment and the price of travel should be a cornerstone of future research efforts. The evidence in chapter 5 confirms that many neighborhood features can be understood within a price framework, and that a price (and consumer demand) framework is essential for illuminating the complex relationship between design and travel. If there is one overarching message of this book, it is that the demand framework presented in chapter 4, and its prominent role for prices (however measured), is essential in understanding the link between urban design and travel.

Beyond that basic argument, several questions emerge. Geographic scale has been overlooked; future analysts should give more attention to the distances over which any link between urban design and travel is evident. Possibly even more important is the extent to which statistical results are biased by the link between residential location choice and preferences about travel. Those prefering alternatives to automobile travel can be expected to select into neighborhoods that offer automobile alternatives, either by buying or renting residences in those neighborhoods. Thus, simply comparing travel patterns in neighborhoods with different design characteristics—the dominant mode of empirical research on this topic—will give biased estimates of the influence of neighborhood design on travel. Our work addressing this issue is only a start.

We also repeat that the policy prescriptions from analyses of urban design and travel hinge crucially on one's assessment of the supply of those neighborhoods. If the supply of one sort of neighborhood is artificially constrained, say, through government regulation, then building more of those neighborhoods could potentially influence travel behavior even if that only provides places for persons who want to live there but who previously could not afford to. Hence, it would be very useful to examine the incentives affecting how local governments regulate and otherwise influence land use, and in turn whether those regulations unduly restrict particular neighborhood types. This work remains primitive and thus there is much to learn.

City building will always be linked to transportation, and vice versa, and the new design and planning strategies are, in many respects, brimming with promise. Yet understanding how those designs influence travel, how they might be implemented, and how to evaluate their transportation goals is a fundamental and still incomplete task.

Any urban design strategy should be encouraged in those respects in which it succeeds. Regarding traffic solutions, our goal here was to clarify what we know while also suggesting how we can learn more. In the end travel remains a very complex story. Travel by design remains an exciting challenge.

Appendix

Data and Data Collection Methods for Chapter 7, by Source

Transit Authorities

As described in chapter 7, there are five transportation authorities that operate or are planning passenger rail lines in southern California: MTA (Metropolitan Transportation Authority, Los Angeles), OCTA (Orange County Transportation Authority, Orange County), NCTD (North County Transit District, northern San Diego County), MTDB (Metropolitan Transit Development Board, central and southern San Diego County), and SCRRA (Southern California Regional Rail Authority, MetroLink). Each of these authorities were contacted to request rail line maps, station addresses, and dates at which time stations would become operational.

Stations

Many of the most recent rail transit studies indicate that the sphere of influence on adjacent development for light rail transit (LRT) stations is approximately one quarter mile in radius (e.g., Bernick and Carroll, 1991; Bernick and Hall, 1992; Cervero, 1994c). The *Thomas Guide Street Guide and Directory* (1994) was used to locate the 232 proposed or existing rail transit stations in southern California and identify jurisdictions. Although the half-mile circle centered on a station was often within one municipality, the area for some stations included up to three separate jurisdictions. Eighty jurisdictions were identified as being within the quarter mile radius of existing or proposed rail transit stations in southern California. Once the initial identification was made, each jurisdiction was phoned and a request was made for appropriate and recent zoning maps.

Zoning: Categories and Measurements

We organized zoning data within one quarter mile of each transit station into six categories. All cities organize their zoning into more precise categories, but for our purposes, we use only the six listed below. These six categories allow us to compare land-use data between the target jurisdictions by creating uniform categories that apply to land use in each jurisdiction.

Low- to Medium-Density Residential: less than or equal to 15 dwelling units (d.u.s) per acre.

High Density Residential: greater than or equal to 15 d.u.s per acre.

Commercial: all commercial and office professional, not including heavy commercial zoning.

Mixed Use: any area where commercial and residential uses may occur simultaneously.

Industrial/Manufacturing: industrial, manufacturing, heavy commercial and any other commercial/industrial zoning classifications. (Note: Land within the zoning categories of heavy commercial and commercial/industrial is included in our industrial/manufacturing category. Commercial uses in most areas that are zoned heavy commercial or commercial/industrial are wholesale warehouses. Sales tax revenues are generally collected at the point of sale and not at the distribution center, thus these warehouses do not typically generate sales tax revenues for their city of residence. From a municipality's perspective, the fiscal and economic characteristics of warehouses are more likely to be similar to industrial/manufacturing land uses than to commercial land uses.)

Other: including open space, rights of way, government properties, public properties, waterways, streets and highways, and unzoned areas.

It was problematic to categorize residential land use according to densities across multiple jurisdictions. Most jurisdictions apply the terms high, medium, or low density, or combinations of the three, to residential zoning. A lesser number of cities combine the aforementioned terms with the terms single family, two family, or multifamily to categorize residential properties. The densities attached to these terms can vary greatly, especially between urban and rural locations. In urban locations, high-density zoning may allow 60–70 d.u.s per acre, while in rural areas high-density zoning may only allow 10–12 d.u.s per acre.

For most jurisdictions, the categorization of residential land can be characterized as follows: estate density (0–2 d.u./ac.), low density (3–4 d.u./ac.), medium density (4–8 d.u./ac.), medium-high density (8–14 d.u./ac.), and high density (more than 15 d.u./ac.). In jurisdictions that use single-family, two-family, and multifamily zoning categories, densities are generally less than 8 d.u. per acre in single family, less than 15 d.u. per acre in two family, and greater than 15 d.u. per acre in areas zoned for multifamily residential.

The density range in the residential classifications in the thirty-five municipalities in Table A.1 do not match the previously mentioned general classifications. (These exceptions have been assigned to either the "low- to medium-density" or "high-density" category based on information received from each municipality on the average densities of each zoning classification. If the average density in the classification is below 15 d.u./ac., that classification is included in the low-density category, and if the average density of the area is above 15 d.u./ac., the area is included in the high-density category.)

Table A.1: High-Density Zoning Definitions, for Local Governments That Are Exceptions to General Classifications in Appendix A

Municipality/County	High Density Designation (dwelling units/acre)
Baldwin Park	Multifamily 12.1–20
Brea	High density 9.7–24.9
Carson	Multifamily 8–25
Cerritos	Medium density >15.5
Commerce	Medium-density multifamily 0–27, high-density multifamily >21.78
Costa Mesa	High density 13–20
Covina	Multifamily >14.6
Downey	Two family 9–17, multifamily 18–24
El Segundo	Two family >15, multifamily >33.9
Escondido	Medium multifamily 16–22
Hawthorne	Medium density 8.1–17, high density 17.1–40
Huntington Beach	Multifamily townhouse 14.7, multifamily apartment 14.52–21.78
Irvine	Medium-high density 10–25, high density 25–40
Laguna Niguel	No density range, all specific plan projects.
Loma Linda	High density 9.1–13, very high density 13.1–20
Lynwood	Multifamily 14.1–18
Mission Viejo	High density 6.5–14
National City	Two family 17.4, multifamily >22.8
Ontario	Medium density 16, high density 25
Orange	Medium-low density 6–15, medium-high density 15–24
Pasadena	Multifamily 12–48
Rancho Cucamonga	Medium-high density 14–24
Redlands	Medium density <17.4
Redondo Beach	Medium-high density >15
Rialto	Multifamily 13–21
San Bernardino	Medium-high density 24, high density 36
San Clemente	Medium density >15
San Diego City	High density >14.5
Santa Clarita	Medium-high density 15.1–25
Santa Fe Springs	Multifamily townhouse 14.7, multifamily apartment 20.7
Santee	Medium-high density 14–22, high density >22
Simi Valley	Medium-high density 8–16, high density >16
Solana Beach	High density 13–20
Upland	Multifamily 9.9–30
Vista	Multifamily 6.6–21.8

Notes

CHAPTER 1

1. Some might object that the early automobile transportation planners very much sought to change urban form to accommodate the automobile and also at times sought to solve societal problems by providing a new mode of transportation. Certainly, there was some utopianism surrounding the automobile in the early decades of the twentieth century, and some planners did advocate that cities could be beneficially "transformed" by providing for better automobile travel (see, e.g., Altshuler, 1965; Foster, 1981; Rose, 1990, esp. chapter 1). Yet those planners were often not dominant in debates that usually deferred to road engineers, and most transportation planning of the first half of the twentieth century sought to respond to the rapidly increasing demand for automobile travel. This is importantly different from the more recent goal of using urban design to, among other things, change how people wish to travel. In the former instance, even if planners sought to transform cities and transportation systems, the key players viewed themselves as responding to changes in travel behavior that were caused by other factors, such as the popularity of the newly developed automobile. In the latter case of the more recent urban designs, the projects often have the more ambitious behavioral goal of seeking to change the desired travel patterns of some persons.

2. Other recent discussions of their work are found in Abrams (1986), Boles (1989), Bookout (1992b), Dunlop (1989, 1991), Kelbaugh (1989), Knack (1989), Leccese (1990), Mahoney and Easterling (1991), Rowe (1991), Carson, Wormser and Ulberg (1995), and Ryan and McNally (1995).

3. The mixed views the architectural profession has held toward the suburbs, ranging from disdain to merely aesthetic, are perhaps part of the story. See the discussion in Boles (1989).

4. Except that, as Calthorpe (1993) emphasizes, traditional small towns tend to lack the densities required to support transit. Fink (1993) also argues that the Neotraditional model, based in many ways on the prototypical Eastern small town, does not apply well to the more decentralized character of the western United States.

5. Interestingly, Duany (1989) emphasizes that these communities are not typically permitted under standard building and planning codes. A central feature of his firm's town plans has been their codes, which provide both for more flexibility in some respects, such as allowing narrower streets, and for less in other respects, such as prescribing design guidelines for individual

structures. Clear descriptions of how a neighborhood and a planning department might change street codes to benefit existing neighborhoods are found, respectively, in Appleyard (1981) and Fernandez (1994).

6. A sampling includes Calthorpe (1993), Bookout (1992a), Duany and Plater-Zyberk (1992), Kulash, Anglin, and Marks (1990), Beimborn et al. (1991), Lerner-Lam et al. (1991), and Ryan and McNally (1995). More doubtful assessments are found in Kaplan (1990), Leccese (1990), and Handy (1997).

7. Granted, benefit-cost analysis incorporates certain assumptions that have been questioned by various authors, and an emphasis on weighing benefits and costs might overshadow the important question of who reaps the benefits and who pays the costs. Yet the advantages of benefit-cost analysis as a policy tool are considerable, and when applied carefully the disadvantages are not as severe as critics claim. Furthermore, assessing benefits and costs is a useful organizing tool for analyzing not only efficiency questions but also the equity question of who wins and who loses.

CHAPTER 2

1. This is the "desire-line map" technique described in Altshuler (1965, pp. 26–28), who notes that this and similar methods were highly influential in highway siting decisions of the 1940s and 1950s.

2. See, for example, Altshuler's (1965, pp. 17–83) description of the planning for the Intercity Freeway connecting St. Paul and Minneapolis, Minnesota. The siting decision was most heavily influenced by highway engineers who were guided by traffic and cost criteria. While other actors, including city planners and community groups, argued against a route that would cut through established neighborhoods and displace many residents, their arguments did not prevail.

3. The percentage of all trips by private vehicle is from U.S. Department of Transportation (1997, p. 155, table 7-3).

4. Murphy and Delucchi (1998) review the recent literature on the social costs of motor vehicle use generally in the Unites States and summarize the social cost estimates by category.

5. Recent simulation evidence suggests that, under certain conditions, the expected cost of commuting can be reduced more by reducing variations in travel times than by reducing the average travel time. This raises the possibility that, in addition to average travel times and speeds, drivers might also care about nonrecurring congestion and its effect on the variance of travel times. See Noland (1997).

6. For a more complete discussion of externalities and externality regulation, see Cropper and Oates (1992).

7. A formal analysis of the relative advisability of quantity and price regulation is found in Weitzman (1974).

8. The California Motor Vehicle Pollution Control Board required, in 1961, that by the 1963 model year, new automobiles sold in the state must include crankcase blowby devices that would reduce emissions (Lave and Omenn, 1981, pp. 29–30); South Coast Air Quality Management District, 1997, p. 15).

9. For a description and a critical assessment of the effectiveness of Mexico City's no-drive days, see Eskeland and Feyzioglu (1997) and Levinson and Shetty (1992, p. 32).

10. In federal fiscal year 1996, transportation grants to state and local governments were 34 percent of all federal grants excluding those for health

(mostly Medicaid) and income security. Of the transportation grants, over two-thirds were for the federal aid highway system. Both proportions have been roughly constant, with transportation totaling approximately one-third of all federal grants other than Medicaid and income support since the mid-1980s (U.S. Office of Management and Budget, 1997).

11. Some persons claim that vehicle inspection programs have been generally ineffective and have contributed little to the realized reductions in vehicle emissions (see, e.g., Glazer, Klein, and Lave, 1995).

12. Of the six NAAQS pollutants, lead concentrations are excluded from the PSI.

13. For NO_2, only one urban area (Fort Wayne, Indiana) recorded an increase in atmospheric concentration from 1986 through 1995. Four urban areas (Des Moines, Iowa; Honolulu, Hawaii; Phoenix-Mesa, Arizona; and Tucson, Arizona) recorded an increase in ozone concentration between 1986 and 1995. For PM-10, three metropolitan areas (Fort Lauderdale, Florida; Monroe, Louisiana; and Texarkana, Texas–Texarkana, Arkansas) recorded an increase in concentration from 1986 through 1995. Three metropolitan areas (Fort Wayne, Indiana; Knoxville, Tennessee; and West Palm Beach–Boca Raton, Florida) recorded an increase in SO_2 concentration, based on the second highest 24-hour reading, from 1986 through 1995. Source: U.S. Environmental Protection Agency (1996, table A-17).

14. Houston's peak ozone measurement for the year was 0.26 parts per million, which exceeded the highest ozone reading anywhere in the Los Angeles air basin for the 1996 smog season, which ended in October of 1996 (Cone, 1996).

15. The average vehicle miles traveled per household increased from 34 to 41.40 from 1969 to 1990 (U.S. Department of Transportation, 1997, table 7-1, p. 151). The U.S. Energy Information Administration has data on gasoline prices going back to 1977. In inflation-adjusted dollars, motor vehicle gasoline, including taxes, was cheaper in 1996 than in 1977 (Energy Information Administration, 1997).

16. For more detailed discussions of the SCAQMD's Regulation XV program, see Bae (1993), Dill (1998), and Giuliano, Hwang, and Wachs (1993).

17. Note that accidents are not externalities to the extent that well-functioning and complete insurance markets exist. Conversely, the external costs of accidents are those that cannot be compensated by existing insurance markets.

18. For example, Baldassare (1991) states that, in Orange County, California, in 1989, "traffic and transportation" was ranked as the "most serious problem" by 40 percent of survey respondents.

19. The Los Angeles commute time data in Gordon and Richardson (1993) were for all modes, not just automobile travel.

20. See also Levinson and Kumar (1994) for a similar argument. Cervero and Wu (1997, 1998) offer somewhat contrary evidence for the San Francisco area.

21. One example of a "doomsday" analysis is the 1988 California Assembly Office of Research Report that cited predictions of travel speeds as low as seven miles per hour during peak periods on some Los Angeles freeways by the year 2000. That analysis ignored or downplayed behavioral responses, including urban decentralization, peak spreading, telecommuting, and other adaptations that historically have held commute times in check.

22. Giuliano, Hwang, and Wachs (1993) found that average vehicle ridership (which is a ratio of the number of employees arriving at a regulated em-

ployment site divided by the number of cars arriving at the site) increased by 2.7 percent in the first year of the Regulation XV program. They concluded that the increase could be attributed to Regulation XV travel demand management (TDM) plans, but they noted that virtually the entire increase in average vehicle ridership (AVR) was due to changes in carpooling. Giuliano, Hwang, and Wachs questioned whether the first year's increase in AVR could be sustained. See also the discussion in Wachs (1993a).

23. For a discussion of the political opposition to congestion pricing, see, for example, Giuliano (1992), Small, Winston, and Evans (1989), and Wachs (1994). For a summary of congestion pricing experiments throughout the world, see Gómez-Ibáñez and Small (1994).

24. As stipulated in the CPTC's franchise, carpools with three or more persons traveled free for the first two years that the lanes were open. Since January 1, 1998, the CPTC has charged a reduced toll for cars with three or more persons (Sullivan and El Harake, 1998).

25. The private franchise and the idea of charging any toll, as opposed to time varying tolls, were both challenged in the state legislature (Weintraub, 1987), and the project was also challenged by a neighboring county (Gómez-Ibáñez and Meyer, 1993). Yet the opposition revolved around the use of a private franchise and charging any toll for travel. The congestion pricing innovation of charging tolls based on the time of day and expected congestion levels never met with much opposition (Sullivan and El Harake, 1998).

26. Recent evidence suggests that the propensity to use the State Route 91 toll lanes does not vary much with income (Mastako, Rilett, and Sullivan, 1998; Sullivan and El Harake, 1998).

27. See, for example, Park (1952) for an early and influential statement of this and related ideas.

CHAPTER 3

1. Garreau (1991) offers the best-known evidence that this is a recent phenomenon. An alternative and compelling argument that these patterns date back to at least the immediate post–World War II period is presented by Hise (1997).

2. Other critical discussions of this literature and many of these issues are found in Anderson, Kanaroglou, and Miller (1996), Berman (1996), Burchell et al. (1998), Cervero and Seskin (1995), Crane (1996a, 1998a, 1998b, 1999a), Davis and Seskin (1997), Deka and Giuliano (1998), Dunphy et al. (1997), Ewing (1997b), Gibbs (1997), Handy (1996b, 1997), Jones and Breinholt (1993), Moore and Thorsnes (1994), Pickrell (1999), Ryan and McNally (1995), and Wachs (1990). This review does not cover all those materials in order to focus on the structure and development of the literature, and its gaps. Other studies not specifically referenced in this chapter that nonetheless were useful in its preparation include Atash (1993, 1995), Berechman and Small (1988), Brownstone and Golob (1992), Cambridge Systematics (1994), Cervero (1986, 1989a), Cervero and Radisch (1996), Deakin (1991), Ewing (1994, 1995a, 1995b), Ewing, DeAnna, and Li (1998), Gordon, Kumar, and Richardson (1989a), Holtzclaw (1990), Johnston and Ceerla (1995), Kitamura, Chen, and Pendyala (1997), Koppelman, Bhat, and Schofer (1993), McNally and Kulkarni (1997), Mokhtarian, Raney, and Salomon (1997), Newman and Kenworthy (1989), Nowland and Stewart (1991), Levinson and Kumar (1994, 1995), Ong and Blumenberg (1998), Peng (1997a, 1997b), Pivo, Hess, and Thatte (1995), Pivo, Moudon, and Loewenherz (1992), Pipkin (1995), Pivo et

al. (1992), Pushkarev and Zupan (1977), Plane (1995), Roberts and Wood (1992), Southworth (1997), Southworth and Ben-Joseph (1997), Spillar and Rutherford (1990), Steiner (1994), Schimek (1998), Thompson and Frank (1995), and Wachs et al., (1993).

3. Other examples of this approach that address other issues besides the street configuration include Johnston and Ceerla (1995), McNally and Kulkarni (1997), and Kitamura, Chen, and Pendyala (1997).

4. This model and the alternatives are described in more detail in 1000 Friends (1996) and in several other of the LUTRAQ reports. The discussion here is only intended to highlight a few of the many issues and results examined there.

5. Three other literatures highly relevant to the study of these issues are not examined in any detail in this chapter. These concern "travel accessibility," "recreation demand," and "parking." The first emphasizes the measure of proximity and opportunity, among other things, sometimes as distinguished from "mobility." While it traditionally attempts to measure the built environment, this literature increasingly also includes measures of travel demand (e.g., Hansen, 1959; Wachs and Kumagai, 1973; Hanson and Schwab, 1987; Crane and Daniere, 1996; Handy and Niemeier, 1997; Crane and van Hengel, 1998). Alternatively, the study of recreation demand often uses travel to recreational sites to measure the value of those sites. Thus, as in the discussion of behavioral models below, they may use trip length as a measure of the cost of recreational travel. Estimation issues, the endogeneity of trip length, heterogeneity of preferences, and the choices of where to go and how to value those options are examined closely in this work (e.g., Yen and Adamowicz, 1994; Englin and Shonkwiler, 1995; Parsons and Kealy, 1995; Haab and Hicks, 1997; Morey and Waldman, 1998; Train, 1998).

Finally, we do not discuss parking either as a design element and or as a (potential) travel cost, though it is both. The impacts of parking on travel behavior have been explored by Brown, Hess, and Shoup (1998), Shoup (1997), Topp (1993), and Willson (1992, 1995), among others. One theme these authors explore is that parking is normally free to the driver, and thus represents an often substantial subsidy to driving. Shoup (1999) suggests that land use authorities do not recognize the extent of the subsidy in the first place, distorting their own planning decisions as well. Both are relevant to the themes of this book and deserve more attention.

6. Holtzclaw defines "pedestrian access" as (fraction of through streets) × (fraction of roadway below 5 percent grade) × (0.33)/(fraction of blocks with walks) + (building entry setback) + (fraction of streets with controlled traffic).

7. While these studies try to explain pedestrian travel, note that the resident value of pedestrian-friendly environments likely extends beyond travel considerations. Handy, Clifton, and Fisher (1998) found evidence that nonwalkers often placed significant value in having pedestrian-oriented features within reach, implying that these features may be prized as neighborhood amenities and opportunities, even where they are not much utilized.

8. Handy's (1996a) evidence on an inverse relationship between trip distance and trip rates is for supermarket shopping trips by all modes, while Holtzclaw (1994) and 1000 Friends (1993) examined only car trips.

9. Compare this argument with the results of Tertoolen, van Kreveld, and Verstraten (1998) that when confronted with differences in their attitudes toward driving and their actual behavior, the surveyed residents of Gouda, The Netherlands, tended to change their attitudes rather than their driving.

10. Differences in travel behavior by gender have also been linked explic-

itly to urban form issues, though seldom in a demand framework. See, for example, Madden and White (1980), Gordon, Kumar, and Richardson (1989b), Madden and Lic (1990), and Rosenbloom (1993).

11. This is similar to one of the empirical specifications we report in chapter 5, making Kockelman's (1998) approach similar to our own. Both are initial attempts to build stronger links between demand theory and questions concerning urban form and travel. While Kockelman only reports models that use urban form as a determinant of trip length, we also formulate and test alternative specifications that include urban form variables directly in a trip generation regression.

CHAPTER 4

1. Access has been measured in many ways, but is often used to capture scale as well as distance (Handy 1992; Crane and van Hengel, 1998). The number and diversity of potential destinations within some specified distance, such as the number of grocery stores and restaurants, is a typical measure (Hanson and Schwab, 1987). In practice, node composition as well as the spatial distribution of nodes thus both matter. To keep the basic story straightforward, this article abstracts from all aspects of access but linear distance. However, although increasing the diversity of destinations clearly affects the attractiveness of any travel mode for any given travel distance, it does not qualitatively affect the logic of the argument.

2. Handy (1992) and Crane (1996a, 1996b) are the only sources we are aware of that explicitly note this consequence of reducing trip length.

3. Extending the story to allow for more travel modes, such as transit and bicycling, would complicate the narrative and analytics without changing the qualitative nature of the results.

4. Here we employ the term "demand curve" somewhat differently from its usual usage, as it gives the preferred mode corresponding to the total cost of an entire trip, not the number of trips or the trip length per unit cost.

5. The formal statement of the maximization problem should properly include certain conditions on the form of preferences, price-taking behavior, and optimization over other consumption (e.g., see Kreps, 1990). It is assumed the standard and necessary conditions hold.

6. Comparative statics is perhaps the most powerful and certainly the most popular tool in microeconomics. It permits the analyst to ask various "what if" questions, and derive the qualitative answers in some detail. Moreover, the basis for those answers follow transparently from the structure of the model. Though we have glossed over many details in this summary of the method, a fuller treatment would show this is not a "black box" approach to explaining outcomes.

7. It seems likely that work trips are relatively price inelastic, at least in the short run when locational changes are not possible, and that nonwork trips are somewhat more sensitive to travel costs. This is borne out by the mode-choice literature (e.g., Train, 1986). The ANOVA results of Ewing, DeAnna, and Li (1998) indicate that trip generation rates by all modes are inelastic with respect to certain land-use variables, including density, land-use mix, and gravity accessibility measures. Unit trip cost is not explicitly included as an explanatory variable, however.

8. The earlier literature on jobs-housing balancing focused almost exclusively on commuting. Many of the criticisms of jobs-housing balancing (see, e.g., Giuliano, 1992) can be viewed as problems associated with using land

use to influence work trips. Persons might choose to live far from their job for many reasons, such as the availability of local amenities, a preference for living in certain communities, or the high costs of both moving and changing jobs. When the focus changes to nonwork travel, many of those concerns are less vexing.

9. This simplifies matters because we need not necessarily understand all there is to know about nonwork travel to be able to test the link to urban design. Instead, we merely need a framework that isolates the hypothesized link between design and travel in a way that can be tested with available data.

10. Examples include Cervero and Gorham (1995), Kitamura Mokhtarian, and Laidet (1997), Crane and Crepeau (1998), Cervero and Kockelman (1997), and Frank and Pivo (1995), all of whom measure land-use characteristics for geographic areas that correspond to census tracts or smaller units of geography. An exception is Handy (1993), who used gravity measures to examine the effect of accessibility to neighborhood and to more distant regional destinations.

11. As an example, for the Orange County/Los Angeles data used in chapter 5, the census tracts for residents average less than six square miles in area. If tracts were circles (a simplification, because tracts are irregularly shaped), the average tract radius for these data would be 1.34 miles. Nonwork trips for the persons sampled averaged 10.24 miles.

12. For the Orange County/Los Angeles area data, 96 percent of all first trips during the travel diary time period were by private automobile. For the San Diego data, 94 percent of all first trips during the travel diary time period were by private automobile.

13. Since all travel diary respondents are from the greater Los Angeles or San Diego areas, we assume that there are no important variations in fuel cost across persons in our sample.

14. Note that, of the studies reviewed earlier, only Kitamura Mokhtarian, and Laidet al. (1997) give any attention to the need to control for how the value of time spent driving changes as income levels change. We include income quadratically in regressions for the Orange County/Los Angeles data alone, as individual income is available only by broad categories for San Diego.

15. Yet land use and urban design characteristics might be endogenous to travel behavior if persons choose, for example, residence locations based in part on how they wish to travel. This problem, and a proposed solution, is discussed below.

16. We experimented some, in earlier work, with models that treated median trip distance and speed as endogenous variables. We were not able to identify good instruments for median distance and speed, and so were not satisfied with the performance of those models. The significance of the land-use variables was generally not affected by the choice of whether or not to use instrumental variables for median trip distance and speed. See Crane and Crepeau (1998) for more discussion.

CHAPTER 5

1. The results presented below draw on the research reported in Boarnet and Sarmiento (1998) and Crane and Crepeau (1998), but this chapter extends that work in several ways as discussed below.

2. The Panel Study of Southern California Commuters is described more fully in Brownstone and Golob (1992). The travel diary data are from the

ninth wave of a ten-wave panel survey. Because persons sometimes stop responding to panel studies, later waves are often not representative samples of either the initial respondents or the underlying population. Boarnet and Sarmiento (1998) investigated whether panel attrition resulted in sample selection bias in the context of trip generation models such as those reported here. They found that the attrition process is independent of the trip generation behavior studied here, and so does not bias either their results or the similar specifications reported here.

3. Our data were collected only for the respondent, typically an employed adult within the household. To adjust for characteristics of the household that might influence individual travel, we included in the model the number of children and the number of household members who work.

4. It is possible that long-distance commuters make more nonwork trips than short-distance commuters during their days off. Thus, we include both the work-day and long-commute dummy variables and the interaction term for those variables to control for the possibility that commute distance affects nonwork trip making on both work days and days off.

5. This is based on 1994 census TIGER (Topologically Integrated Geographic Encoding and Referencing) street maps. One motivation for %GRID is research by Kulkarni (1996), who found that the number of four-way intersections was one of the most influential variables in a cluster analysis used to group Orange County, California, neighborhoods based on how closely those neighborhoods reflected Neotraditional design tenets. Cervero and Gorham (1995) similarly classified neighborhoods as transit- or auto-oriented based in part on an assessment of the percentage of four-way intersections in the neighborhood. Note that %GRID was not constructed by counting four-way and total intersections. That was not possible given the large number of residence locations in this study and limitations of the GIS software. Instead, for each quarter-mile-radius area, we marked the area that contained four-way intersections, and that area was measured with a digital planimeter. We believe that this technique gives measures of the street network that are similar to those that would be obtained by counting intersections.

6. The data for ZIP code population density are from the 1990 census. Retail and service employment densities for ZIP codes, obtained from the Southern California Association of Governments, are for 1994. The data and variables discussed here were first collected in Boarnet and Sarmiento (1998). We are grateful to Sharon Sarmiento and to the research assistants acknowledged in Boarnet and Sarmiento (1998) for assistance in collecting these data.

7. The Orange County/Los Angeles residents also specified income by categories, but a linear income variable was imputed from the categorical income data for Orange County/Los Angeles. See Sarmiento (1995) for details.

8. In the regressions for San Diego, we examine land-use variables characterizing only the respondent's home census tract, with the exception of the street grid variables that are measured within a half mile of each respondent's household.

9. Consider a hypothetical situation: A particular parcel of land may be zoned commercial but is vacant. The land-use data here reflect the vacant status of the land, while the zoning information cannot capture this information. This is an important distinction to make when dealing with issues of transportation and urban form.

10. This sign and significance pattern on the sociodemographic control variables were robust (with minor changes) across virtually all specifications

tested in this research. In many specifications, both income coefficients were significant at the 5 percent level and with opposite signs, suggesting the quadratic effect discussed in chapter 4. The regressions in columns A and B of table 5.4 were also estimated by ordered logit, and the results were not meaningfully different. The results in columns A and B of table 5.4 also do not vary when the variable for cars per drivers is omitted from the model, and similarly the results do not differ when we control for the length of the work commute. See Boarnet and Sarmiento (1998) for more discussion of the robustness of the specification and results.

11. For both San Diego and Orange County/Los Angeles, data were available that allowed us to calculate travel speeds and distances for nonwork automobile trips. The median nonwork car trip speed and distance for each individual (Orange County/Los Angeles) or household (San Diego) are the trip price (or time-cost) variables included in column C of table 5.4 and in similar regression models in later tables.

12. These results are consistent across virtually all of the specifications tested. Using ordered logit produced no meaningful differences in the results. See Crane and Crepeau (1998) for ordered logit results for similar specifications.

13. Note that it need not necessarily be the case that commercial concentrations reduce nonwork trip speeds. The data suggest that occurred in the San Diego sample, but results in other urban areas could be different.

14. If urban design provides residential locations for persons who already prefer alternatives to the automobile, the policy questions revolve more around neighborhood development than around driving behavior. One important policy question, in this scenario, is whether neighborhoods that support alternatives to the automobile are undersupplied relative to consumer demand for those types of residences. If so, it becomes necessary to disentangle travel demand questions from issues of the supply of neighborhood types. In this chapter we focus on questions of travel demand. We introduce issues of the supply of neighborhoods with new urban design attributes in chapter 6.

15. Formally, the difficulty is that $E(\mathbf{L}'\mathbf{u}) \neq 0$ and plim $(\mathbf{L}'\mathbf{u}/N) \neq 0$. A similar point holds for the discrete choice model if **L** is correlated with the error term in a discrete choice regression based on equation (7) of chapter 4, but the solution becomes somewhat more complex. For related discussions, see Train (1986, pp. 82–90) or Small and Hsiao (1985).

16. The instrumental variables technique described below is in some ways an attempt to control for an individual's nested choice of, first, residential location, and then trip generation. Train (1986) discusses how nested choice models can be collapsed into instrumental variables models. When, as is the case here, the concern is more with the final choice (trip generation) than the sequence of decisions, an instrumental variables technique can control for confounding influences from earlier steps in the choice process. Of course, a more complete nested choice model of residential location and trip generation also warrants examination as a topic for future research.

17. For an application, see, for example, Levine (1998).

18. For a more detailed discussion of the variables which influence residential location choice and the allowable instruments for land-use variables, see Boarnet and Sarmiento (1998).

19. We test this assumption of instrument exogeneity using overidentification tests, and the results are reported below.

20. At each step in the analysis, we used the instrumental variables technique for regression models that both included and did not include the trip

cost (median speed and median distance) variables. We only report the results that do not include the median speed and distance variables. In most instances, including median speed and median distance produced no change in the sign and significance of the land-use variables.

21. The land-use variables are entered singly or, at most, two at a time, in the regressions in table 5.10. That is done because the overidentification test, described below, requires more instruments than land-use variables. With a total of four land use variables and four instruments, this requires that less than all four land-use variables be used in a single regression. Because RETDEN and SERVDEN are intended to jointly proxy for land-use mix in census tracts, they were included together in column C of table 5.10. The other land-use variables (%GRID and POPDEN) were entered singly in the regressions in table 5.10.

22. The test statistic is equal to the number of observations times the R^2 from a regression of the two-stage least squares residuals from the second stage equation on all included and excluded instruments. The overidentification test is actually a joint test of the exogeneity of the instruments and the appropriateness of the specification. For a description of the test and an application, see Angrist and Krueger (1989, 1994).

23. In column A of table 5.11, the GRID and MIXED land-use variables are both included in the model, because they jointly describe they street grid within a quarter mile of each individual's residence. Similarly, D_CBD and D_CBD2 are both in the regression in column C. For all other columns in table 5.11, land-use variables are entered singly into the regression model. The three variables %RESID, %COMM, and %VACANT jointly describe the land-use mix within the census tract of residence, but because those variables are potentially collinear, they are reported singly in table 5.11. The results do not meaningfully differ when the %RESID, %COMM, and %VACANT variables are included as a group in an instrumental variables regression. The variables for median nonwork car trip speed and distance are omitted in all regressions in table 5.11. When median speed and distance are included in the instrumental variables regressions (not reported here), the coefficients on GRID and %COMM become insignificant.

24. The overidentification statistic rejects the hypothesis of valid instruments, at the 5 percent level, for GRID and MIXED in column A, HEAVY in column B, %RESID in column D, and %VACANT in column F.

25. The overidentification test is a joint test of the instruments and the specification (Johnston and Dinardo, 1996). The instrumental variables routines in table 5.11 are reduced forms of residential location and nonwork travel choices that could be modeled as nested discrete choices. It is possible that the overidentification tests reported in tables 5.10–5.12 are evidence that the instrumental variables specification is inadequate for this problem. An alternative for future work would be to explore more complete nested probit or logit specifications of individuals' choices of residential locations and automobile trip frequencies.

CHAPTER 6

1. There is some evidence that this occurs. Cervero (1994d) reports that, among residents of transit-oriented developments who commuted by rail, 42.5% stated that they commuted by public transit before they moved to a transit-oriented development.

2. By influencing the nature of the urban form, new urban designs can also affect nonresidential location choices, such as job locations and the location of shopping, school, and entertainment destinations. We simplify here by focusing only on residential locations.

3. This is not to say that the scarcity of housing for certain income groups is not a market or social problem. Clearly in many instances it may be. Recent evidence on this question is discussed in Crane (1999b).

4. Even absent any market failure or government regulation that constrains the supply of the new urban designs, one could argue for policy intervention based on equity arguments. The argument might be that policy should seek to increase the availability of homes in New Urbanist neighborhoods. Yet while housing itself might be an important target of such equity policies, we know of no compelling argument why providing a specific housing type (e.g., housing in New Urbanist neighborhoods) is an appropriate equity issue. For that reason, we focus on the question of either market failures or government policies which might constrain the supply of the new urban designs.

5. This is known as the "product diversity" problem in industrial organization (IO). See any modern IO text for a discussion (e.g., Tirole, 1988; Carlton and Perloff, 1994).

6. There is one, possibly serious, exception that we are aware of to our conclusion that there are no important market failures that might constrain the supply of neighborhood types in ways that restrict the supply of the new urban designs. Owner-occupied housing is, for most people, both a residence and the largest component of their personal wealth. Housing, then, is both a consumption good and an investment. To the extent that housing is an investment, persons will care about how easily they can sell a house. This means that even persons with idiosyncratic tastes in homes and/or neighborhoods might see value in buying residences that are more conventional. The more conventional homes in more conventional neighborhoods can be more easily sold. Thus, the investment character of housing could be a factor that constrains the diversity of the demand for house and neighborhood types relative to what persons might desire if housing were only a consumption good. That constraint on diversity in demand might in turn constrain the supply of neighborhood types. Having said that, we see no clear policy solution (for an alternative view, and a proposed policy solution, see Brueckner, 1997a). For that reason, we focus on government regulations that, we believe, point more clearly to suggested policies.

7. A separate important question is how land use regulations and other local public policies, such as finance, affect development patterns to begin with. We implicitly assume in this discussion that the latter follow directly from the former, but there is little systematic evidence to support this. Every practicing planner knows that zoning regulations and general plans can be, and often are, quite different from actual development, for example. Gyourko and Voith (1999) and Pendall (1999) investigate this issue.

8. This can occur even if we judge the city's self-interested motives to be legitimate. For example, in fiscal zoning, one might concede that residents sometimes seek to keep the per capita local tax base high and per capita expenditures low, while noting that those motives disadvantage potential inmigrants in ways that exceed the benefit to city residents. In other instances, one should deny the legitimacy of city motives, such as in cases in which the exclusionary motive is a desire to keep out persons of a particular race. In the latter case, the argument against exclusion becomes even stronger.

9. There is a large literature that examines whether self-interested municipal governments will behave in ways that further social welfare. This literature dates to Tiebout's (1956) seminal article that hypothesized that local governments that compete for residents will, given certain rather restrictive assumptions, efficiently provide local public goods. (Efficiency in this instance implies that local public goods are provided in a fashion that equates, at the margin, the social benefit of providing an extra unit of the public good with the extra cost of providing the public good. This would mimic the efficiency of competitive markets for private goods, with the exception that the marginal social benefit for a public good is the sum of the marginal social benefit that accrues to all individuals who consume the good. On this last point, see Samuelson [1954].) The articles that followed Tiebout's work were often attempts to analyze whether competing local governments will provide the diversity of public goods that is desired by consumers. To the extent that neighborhood type is a public good, it can be analyzed within a Tiebout framework. We do not do that here largely because of the analytical and conceptual difficulties that are often encountered in attempting to formalize or draw policy insight from Tiebout theory. Instead, we focus on the behavioral details of one specific class of incentives, namely, fiscal and economic incentives.

10. For an account of how concerns about the intrametropolitan economic impacts of transportation shaped the transportation politics of one urban area, see Adler (1991).

11. This point about sales tax finance applies most directly to California, which is the source for our empirical evidence. In California, a portion of locally generated sales taxes is rebated back to municipal governments by the state. For that reason, municipal governments in California have incentives to increase the number of taxable transactions that occur within their borders (for discussion, see Fischel, 1989; Fulton, 1996; Lewis and Barbour, 1999). Yet as other states pursue forms of local sales tax finance, the growing fiscal pressures to favor commercial over residential development will extend beyond California.

12. Of course, regulations that constrain a particular kind of neighborhood might have other benefits. For example, traditional zoning codes separate residential from commercial and industrial land uses based on the long-popular notion that residents benefit from a physical separation of those land uses. Thus, any move to lessen that zoning regulation must account for the benefits provided by the regulation itself. In short, one should balance any social benefits due to regulation against the cost of, for example, constraining the supply of mixed-use developments. In the discussion below, we often focus only on the constraints imposed by local government regulations, but policy analysts should also be cognizant of any benefits due to zoning or similar regulations.

13. For similar and related arguments, see Downs (1992), especially chapter 6 and appendices C and D.

CHAPTER 7

1. The discussion and empirical research presented in this chapter are drawn from Boarnet and Crane (1997, 1998a, 1998b). In particular, many of the findings in this chapter were first published in Boarnet and Crane (1997, 1998a), and we thank the *Journal of the American Planning Association* and the *Journal of Planning Education and Research* for permission to reprint portions of that work here.

2. Yet the evidence also shows that many transit-area residents (as many as 42 percent according to Cervero, 1994d) were transit riders before they moved, implying that TOD had a smaller stimulative impact on transit ridership than it may appear.

3. Municipalities can use other tax instruments. There is the local option at the county level to raise the sales tax rate, but in addition to the limited influence that individual municipalities have over the county tax structure, voter support for such increases has lately been limited to funds earmarked for transportation infrastructure. Some localities use special taxes such as business license fees and hotel occupancy taxes, but the sales and property taxes are still large revenue sources for most municipalities.

4. Many Southern California rail lines use existing right-of-way. In Orange County, the right-of-way that could most easily be converted to urban rail was the old Pacific Electric right-of-way that extends from Watts in southeast Los Angeles to Santa Ana. Earlier rail studies had concluded that such a route did not serve the county's growing employment and population centers, and that right-of-way, owned by the Orange County Transportation Commission (OCTC) at the time, was never seriously considered in the latest round of rail transit planning (OCTC, 1980).

5. The station sites used in this chapter are the ones suggested in Orange County Transportation Authority (OCTA, 1991). Since then, the OCTA has embarked on an alternatives analysis that focuses on a mile-wide planning corridor centered around the preferred alignment developed in Central Orange County Fixed Guideway Project (1990) and OCTA (1991).

6. The evidence presented below is drawn from a comprehensive analysis of all rail transit stations in Southern California (specifically Los Angeles, Orange, Riverside, San Bernardino, San Diego, and Ventura counties). One might wonder whether results from southern California are representative of other urban areas. While not able to definitively answer that question, we believe that the intercity economic competition that drives our southern California findings is also important in many other urban areas. An interesting exercise for future research would be to examine the extent to which the trend toward commercial development near rail stations documented below exists in other metropolitan areas with well established rail transit systems.

7. The NCTD lines opened after the empirical analysis described below. At the time of the empirical analysis, all of the NCTD stations were classified as proposed rather than operating.

8. Additional analysis suggested that the results in table 7.3 are not due to large outliers that could skew line averages. On most lines, the majority of stations have commercial station ratios that are larger than residential station ratios. This suggests a tendency toward commercial zoning at most of the stations on almost all of the rail transit lines in Southern California. See Boarnet and Crane (1997) for details.

9. Given that rail transit systems often create small changes in regionwide accessibility, those lines would be expected to induce small changes in land use (e.g., Meyer and Gómez-Ibáñez, 1981; Giuliano 1989). Others have noted that land-use responses, if they occur, often follow the inauguration of rail transit service or other interventions by several years (e.g., Knight and Trygg, 1977; Wachs, 1993a, 1993b).

10. The ideal way to examine this question would be to look at changes in the zoning code after rail transit service was established at the various stations. Such changes in zoning could more easily be associated with the rail transit as opposed to the preexisting right-of-way. However, collecting zoning

data at two different points in time proved to be prohibitively expensive both in terms of the project resources and in terms of analytical clarity. Different municipalities keep historical zoning records at different levels of detail, and the codes changed at different times in different cities. In many southern California communities, historical zoning records are difficult or impossible to obtain. As an example, DiMento, van Hengel, and Ryan (1997), after considerable effort, were able to obtain pre-1992 zoning for only eight of the thirteen jurisdictions spanned by the Los Angeles MTA's Green Line. Thus, the most realistic comparison across cities was to look only at current zoning. The empirical models developed and tested below are in some ways an attempt to control for the unavailability of reliable data on zoning changes.

11. Note that if our hypothesis does not hold, LINESHARE is a nonsense variable, and there would be no systematic relationship between LINESHARE and zoning patterns, commercial station ratios included. Thus a test of whether LINESHARE is consistently significantly positive is a powerful test of our hypothesis.

12. We did not have an a priori expectation about the sign of the coefficient on DENSITY. Dense cities might already have large concentrations of commercial uses, and thus commercial concentrations near stations might look more like the rest of the city, so high DENSITY would be associated with low commercial station ratios. On the other hand, dense cities might be centrally located economic centers themselves, and they might be especially able to establish economic and commercial centers near stations. This could lead not only to large amounts of commercial zoning near stations in dense cities, but also to large commercial station ratios for stations in dense cities.

13. While we had some right-of-way information for lines, we preferred measures that were based on nearby zoning characteristics. That is because we could not determine what effect a particular right-of-way would have on land use. Thus, we preferred to infer the extent to which right-of-way constrained observed zoning patterns, by developing measures based on zoning near rail transit lines.

14. The MTDB South and Centre City lines were excluded in this step because they were known to have used preexisting rights-of-way, rather than because they had high industrial concentrations near their stations. See Boarnet and Crane (1997) for details.

15. See Boarnet and Crane (1997) for a more complete description of the criteria for identifying and excluding right-of-way constrained lines.

16. Recall that to understand zoning patterns in the quarter-mile area around stations, we gathered zoning data for eighty cities. Some stations are close enough to a city border that a quarter-mile circle centered on that station falls into more than one jurisdiction. The fiscal data below are only for the cities that have rail transit stations within their borders.

17. Note that, because the dependent variable is the ratio of the percentage of station area land in commercial divided by the same ratio for the municipality, we have no a priori expectation about the sign of the coefficient on the density variable. Density is included to be certain that the fiscal variables are not proxying for characteristics of a city's urban form. Because Proposition 13 greatly limited the rate at which existing property appreciates (unless that property is sold), but assesses new development at market value, there can be a correlation between the property tax base and the age of the development. Since most new development is in less dense suburban municipalities, density is included in the regression to reduce the possibility that *ptaxdep* proxies for a relation between density and commercial station ratios.

18. The largest part of this gap between sales and property taxes and total revenues is grants from the state government to municipalities. In 1990/1991, such grants comprise, on average, 18 percent of local revenue for the sixty-five cities in our database.

19. Sixteen stations are omitted from this regression because of missing data.

20. See Boarnet and Crane (1998a) for complete details of these tests.

CHAPTER 8

1. Nearly all this chapter was first published as Marlon Boarnet and Nicholas Compin, "Transit-Oriented Development in San Diego County: The Incremental Implementation of a Planning Idea," in the *Journal of the American Planning Association 65*, 1999. We thank Nicholas Compin, for his help and insight into this topic.

2. In some cities, the planning function is housed in either a community development or a development services department, in which case we interviewed the head of that department. Throughout this chapter, the term "planning director" refers to all directors interviewed for this study.

3. Although the interviewer did not use the term "transit-oriented development," each planning director was familiar with the concept and used the term in their discussions.

4. The San Diego Trolley's North Line includes the Santa Fe Depot and the County Center/Little Italy stations, which were initially constructed as an extension of the Trolley's South Line.

5. Since the case study research took place, the MTDB changed the way it refers to trolley routes. Instead of individual segments, the trolley system is now segregated into the Blue and Orange Lines. The Blue Line includes the Mission Valley segment, the Old Town segment, the Centre City segment, and the entire South Line. The Orange Line includes the Bayfront segment and the East segment.

6. The zoning data are discussed in more detail in the appendix.

7. The land near the Pacific Fleet station is owned by the U.S. Navy, and comparable zoning information was not readily available for this station. For the quarter-mile circle around the Harborside station, 47 percent of the land is in the City of San Diego and 53 percent is owned by the Navy. The percentages in table 8.1 only include the land in the City of San Diego. Similarly, 54 percent of the quarter-mile around the 8th Street station is owned by the Navy. The percentages for that station are based on the 46 percent of the land that is in National City. For the quarter-mile area around the San Ysidro station, 43 percent of the land is in Mexico. The zoning data in table 8.1 reflect only the 57 percent of the land in the San Ysidro station's quarter-mile area that is in San Diego.

8. Residential density classifications vary from one community to the next, but for Chula Vista, El Cajon, La Mesa, and Lemon Grove it was possible to identify zoning categories with densities of at least fifteen dwelling units per acre, and for those cities the "multifamily residential" category corresponds to a minimum of fifteen dwelling units per acre. In National City, what we classify as multifamily residential is zoned for densities greater than 17.4 dwelling units per acre; in San Diego, what we identify as multifamily residential is zoned at densities greater than 14.5 dwelling units per acre; and in Santee, what we call multifamily residential is zoned at densities higher than fourteen dwelling units per acre.

9. There are proposed TOD developments in Chula Vista (the Otay Ranch project) and Santee (Santee Civic Square), but construction has not started on either project. Even accounting for the planning activity associated with Otay Ranch and Santee Civic Square, it is still fair to say that the bulk of San Diego County TOD has so far been within the cities of La Mesa and San Diego.

10. In the early 1990s, development plans for America Plaza included a hotel, but recent changes to those plans resulted in the proposed development of a library instead.

11. Some authors have suggested that opposition from current residents is also an obstacle to the development of high-density residential projects near stations (Deakin and Chang, 1992). Planners in this study were not asked about their perception of the public's reaction to high-density residential development in their cities, although the planning director in La Mesa addressed that issue in regards to the Grossmont Specific Plan Area without any prompting from the interviewer. Also note that none of the seven planning directors interviewed in this study cited public opposition to residential development as a factor that influenced their station-area development plans. While this does not prove that resident opposition was unimportant, it suggests that it was not a major factor in the viewpoints of the planning directors interviewed.

12. Redevelopment zones allow cities to use the power of eminent domain to acquire property for private development. The Redevelopment Act also allows communities to create tax-increment financing districts that can issue bonds against future property tax increases (Fulton, 1991, pp. 243–244).

13. Planner's responses and discussion involving TOD were completely without prompting from the interviewer. We did not use the term TOD, nor did we discuss in any other than the most limited sense the idea.

14. For a discussion of policies that can be used to mitigate the fiscal impacts of TOD, see chapter 7 and Boarnet and Crane (1997).

CHAPTER 9

1. Recall that, for the Southern California municipalities studied here, a large part of the story is likely how localities influenced station siting to buttress existing commercial and office centers, rather than how governments influenced development near existing stations. Yet both effects are important, and especially the case study evidence from San Diego documents how local governments are constrained and in turn impose constraints on the developments that they seek or pursue near rail stations.

References

1000 Friends of Oregon. 1993. *The Pedestrian Environment: LUTRAQ Report Volume 4A.* Prepared by Parsons Brinckerhoff Quade & Douglas, Portland, Oregon.

1000 Friends of Oregon. 1996. *Analysis of Alternatives: LUTRAQ Report Volume 5.* Prepared by Cambridge Systematics, Inc. and Parsons Brinckerhoff Quade & Douglas (May), Portland, Oregon.

1000 Friends of Oregon. 1997. *Making the Connection—Technical Report: LUTRAQ Report Volume 8.* Prepared by Parsons Brinckerhoff Quade & Douglas (March), Portland, Oregon.

Abrams, Janet. 1986. The Form of the (American) City, *Lotus International No. 50*: 7–29.

Adler, Sy. 1991. The Transformation of the Pacific Electric Railway: Bradford Snell, Roger Rabbit, and the Politics of Transportation in Los Angeles. *Urban Affairs Quarterly 27,* 1: 51–86.

Alonso, William. 1964. *Location and Land Use.* Cambridge, Mass.: Harvard University Press.

Altshuler, Alan A. 1965. *The City Planning Process.* Ithaca, N.Y.: Cornell University Press.

Altshuler, Alan A., and José A. Gómez-Ibáñez. 1993. *Regulation for Revenue: The Political Economy of Land Use Exactions.* Washington, D.C.: The Brookings Institution and the Lincoln Institute of Land Policy.

Anas, Alex, Richard Arnott, and Kenneth A. Small. 1997. *Urban Spatial Structure.* Working Paper No. 357. Berkeley: University of California Transportation Center (March).

Anderson, William P., Pavlos S. Kanaroglou, and Eric J. Miller. 1996. Urban Form, Energy and the Environment: A Review of Issues, Evidence and Policy. *Urban Studies 33,* 1: 7–35.

Angrist, Joshua, and Alan B. Krueger. 1989. *Why Do World War II Veterans Earn More Than Nonveterans?* National Bureau of Economic Research Working Paper 2991. Cambridge, Mass: National Bureau of Economic Research.

Angrist, Joshua, and Alan B. Krueger. 1994. Why Do World War II Veterans Earn More Than Nonveterans? *Journal of Labor Economics 12,* 1: 74–97.

Appleyard, Donald. 1981. *Livable Streets.* Berkeley: University of California Press.

Atash, Farhad. 1993. Mitigating Traffic Congestion in Suburbs: An Evaluation of Land-Use Strategies. *Transportation Quarterly 47,* 4: 507–524.

Atash, Farhad. 1995. Reorienting Metropolitan Land Use and Transportation Policies in the USA. *Land Use Policy 13,* 1: 37–49.

Bae, Chang-Hee Christine. 1993. Air Quality and Travel Behavior: Untying the Knot. *Journal of the American Planning Association 59,* 1: 65–74.

Banerjee, Tridib, and William C. Baer. 1984. *Beyond the Neighborhood Unit: Residential Environments and Public Policy.* New York: Plenum Press.

Baldassare, Mark. 1991. Transportation in Suburbia: Trends in Attitudes, Behaviors and Policy Preferences in Orange County, California. *Transportation 18*: 207–222.

Beimborn, Edward, Harvey Rabinowitz, Peter Gugliotta, Charles Mrotek, and Shuming Yan. 1991. *The New Suburb: Guidelines for Transit Sensitive Suburban Land Use Design.* Milwaukee: Center for Urban Transportation Studies, University of Wisconsin-Milwaukee.

Ben-Akiva, M., and Steven R. Lerman. 1985. *Discrete Choice Analysis: Theory and Application to Travel Demand.* Cambridge, Mass.: MIT Press.

Ben-Joseph, Eran. 1997a. Changing the Residential Street Scene: Adapting the Shared Street (Woonerf) Concept to the Suburban Environment. *Journal of the American Planning Association 61,* 4: 504–515.

Ben-Joseph, Eran. 1997b. Traffic Calming and the Neotraditional Street. *Resource Papers for the 1997 International Conference.* Washington, D.C.: Institute of Transportation Engineers, 47–52.

Berechman, Joseph and Kenneth Small. 1988. Modeling Land Use and Transportation: An Interpretive Review for Growth Areas. *Environment and Planning A, 20:* 1285–1309.

Berman, Michael Aaron. 1996. The Transportation Effects of Neo-Traditional Development, *Journal of Planning Literature 10,* 4: 347–363.

Bernick, Michael. 1990. *The Promise of California's Rail Transit Lines in the Siting of New Housing.* Report to the California Senate Transportation Committee and the Senate Housing and Urban Affairs Committee.

Bernick, Michael, and Michael Carroll. 1991. *A Study of Housing Built Near Rail Transit Stations: Northern California.* IURD Working Paper 582. Berkeley: University of California Institute of Urban and Regional Development.

Bernick, Michael, and Robert Cervero. 1997. *Transit Villages in the 21st Century.* New York: McGraw-Hill.

Bernick, Michael, and Peter Hall. 1990. *Land Use Law and Policy for Maximizing Use of California's New Inter-Regional Rail Lines.* IURD Working Paper 523. Berkeley: University of California Institute of Urban and Regional Development.

Bernick, Michael, and Peter Hall. 1992. *The New Emphasis on Transit-Based Housing Throughout the United States.* IURD Working Paper 580. Berkeley: University of California Institute of Urban and Regional Development.

Boarnet, Marlon G. 1994a. An Empirical Model of Intrametropolitan Population and Employment Growth. *Papers in Regional Science, The Journal of the RSAI, 73,* 2: 135–152.

Boarnet, Marlon G. 1994b. The Monocentric Model and Employment Location. *Journal of Urban Economics, 36,* 1: 79–97.

Boarnet, Marlon G. 1997a. Highways and Economic Productivity: Interpreting Recent Evidence. *Journal of Planning Literature 11,* 4: 476–486.

Boarnet, Marlon G. 1997b. Infrastructure Services and the Productivity of Public Capital: The Case of Streets and Highways. *National Tax Journal 50,* 1: 39–57.

Boarnet, Marlon G., and Nicholas S. Compin. 1996. *Transit-Oriented Development in San Diego County: Incrementally Implementing a Comprehensive Idea.* Working Paper 1996–47. Irvine, Calif.: University of California, Irvine Department of Urban and Regional Planning.

Boarnet, Marlon G., and Nicholas S. Compin. 1999. Transit-Oriented Development in San Diego County. *Journal of the American Planning Association 65.* 2: 80–95.

Boarnet, Marlon G., and Randall Crane. 1997. L.A. Story: A Reality Check for Transit-Based Housing. *Journal of the American Planning Association 63,* 189–204.

Boarnet, Marlon G., and Randall Crane. 1998a. Public Finance and Transit-Oriented Planning: Evidence from Southern California. *Journal of Planning Education and Research 17,* 3: 206–219.

Boarnet, Marlon G., and Randall Crane. 1998b. *The Influence of Land Use on Travel Behavior: Specification and Estimation Issues.* Working Paper. Berkeley: University of California.

Boarnet, Marlon G., Eugene J. Kim, and Emily Parkany. 1998. Measuring Traffic Congestion. *Transportation Research Record* 1634: 93–99.

Boarnet, Marlon G., and Sharon Sarmiento. 1998. Can Land Use Policy Really Affect Travel Behavior? *Urban Studies 35,* 7: 1155–1169.

Boarnet, Marlon G. 1999. Road Infrastructure, Economic Productivity, and the Need for Highway Finance Reform. *Public Works Management and Policy* 3, 4: 289–303.

Boles, Daralice D. 1989. Reordering the Suburbs. *Progressive Architecture 70* (July): 78–91.

Bollens, Scott A. 1992. State Growth Management: Intergovernmental Frameworks and Policy Objectives. *Journal of the American Planning Association 58,* 4: 454–466.

Bollinger, Christopher R., and Keith R. Ihlanfeldt. 1997. The Impact of Rapid Rail Transit on Economic Development: The Case of Atlanta's MARTA. *Journal of Urban Economics 42,* 179–204.

Bookout, Lloyd W. 1992a. Neotraditional Town Planning: Cars, Pedestrians, and Transit. *Urban Land 51* (February): 10–15.

Bookout, Lloyd W. 1992b. Neotraditional Town Planning: Toward a Blending of Design Approaches. *Urban Land 51* (August): 14–19.

Brown, Jeffrey, Daniel Hess, and Donald C. Shoup. 1998. *Unlimited Access.* Working Paper. Los Angeles: Institute of Transportation Studies, UCLA.

Brownstone, David, and Thomas F. Golob. 1992. The Effectiveness of Ridesharing Incentives: Discrete Choice Models of Commuting in Southern California. *Regional Science and Urban Economics 22,* 1: 5–24.

Brueckner, Jan K. 1997a. Consumption and Investment Motives and the Portfolio Choices of Homeowners. *Journal of Real Estate Finance and Economics 15,* 2: 159–180.

Brueckner, Jan K. 1997b. Infrastructure Financing and Urban Development: The Economics of Impact Fees. *Journal of Public Economics 66,* 3: 383–407.

Burchell, Robert W., and David Listokin. 1980. *The Practitioner's Guide to Fiscal Impact Analysis.* Center for Urban Policy Research, Rutgers University.

Burchell, Robert W., et al. 1998. *Costs of Sprawl Revisited: The Evidence of Sprawl's Negative and Positive Effects.* Washington, D.C.: Transit Cooperative Research Program, Transit Research Board, National Research Council (September).

California Assembly Office of Research. 1988. *California 2000: Gridlock in the Making.* Sacramento: California Assembly Office of Research.

Calthorpe, Peter. 1993. *The Next American Metropolis: Ecology Community and the American Dream.* New York: Princeton Architectural Press.

Cambridge Systematics. 1994. *The Effects of Land Use and Travel Demand Strategies on Commuting Behavior.* Washington, D.C.: Federal Highway Administration.

Carlson, Dan, Lisa Wormser, and Cy Ulberg. 1995. *At Road's End: Transportation and Land Use Choices for Communities.* Washington, D.C.: Island Press.

Carlton, Dennis W., and Jeffry M. Perloff. 1994. *Modern Industrial Organization,* 2nd ed. New York: HarperCollins.

Central Orange County Fixed Guideway Project. 1990. *Prospectus for the COCFGP. Cities of Santa Ana, Anaheim, Costa Mesa, Irvine, and Orange.* Los Angeles: COCFGP (September).

Cervero, Robert. 1986. *Suburban Gridlock.* New Brunswick, N.J.: Center for Urban Policy Research Press.

Cervero, Robert. 1989a. *America's Suburban Centers: The Land Use-Transportation Link.* London: Allen and Unwin.

Cervero, Robert. 1989b. Jobs-Housing Balance and Regional Mobility. *Journal of the American Planning Association 55,* 136–150.

Cervero, Robert. 1993. *Ridership Impacts of Transit-Focused Development in California.* IURD Monograph 45. Berkeley: University of California Institute of Urban and Regional Development.

Cervero, Robert. 1994a. *Transit-Supportive Development in the United States: Experiences and Prospects.* IURD Monograph 46. Berkeley: University of California Institute of Urban and Regional Development.

Cervero, Robert. 1994b. Rail Transit and Joint Development: Land Market Impacts in Washington, DC, and Atlanta. *Journal of the American Planning Association 60,* 1: 83–94.

Cervero, Robert. 1994c. Rail-Oriented Office Development in California: How Successful? *Transportation Quarterly 48,* 1: 33–44.

Cervero, Robert. 1994d. Transit-Based Housing in California: Evidence on Ridership Impacts. *Transport Policy 3,* 174–183.

Cervero, Robert. 1995. *BART@20: Land Use and Development Impacts.* IURD Monograph 49. Berkeley: University of California Institute of Urban and Regional Development.

Cervero, Robert. 1995c. Sustainable New Towns: Stockholm's Rail-Served Satellites. *Cities 12* (February): 41–51.

Cervero, Robert. 1996. Mixed Land-Uses and Commuting: Evidence from the American Housing Survey. *Transportation Research A: Policy and Practice 30,* 5: 361–377.

Cervero, Robert, Michael Bernick, and Jill Gilbert. 1994. *Market Opportunities and Barriers to Transit-Based Development in California.* Working Paper 223. Berkeley, Calif.: University of California Transportation Center.

Cervero, Robert, and Peter Bosselmann. 1994. *An Evaluation of the Market Potential for Transit-Oriented Development Using Visual Simulation Techniques.* IURD Monograph 47. Berkeley, Calif.: University of California Institute of Urban and Regional Development.

Cervero, Robert, and Roger Gorham. 1995. Commuting in Transit Versus Automobile Neighborhoods. *Journal of the American Planning Association 61,* 210–225.

Cervero, Robert, and Kara Kockelman. 1997. Travel Demand and the 3Ds: Density, Diversity, and Design. *Transportation Research D: Transport and Environment 2,* 3: 199–219.

Cervero, Robert, and John Landis. 1995. The Transportation-Land Use Connection Still Matters. *Access 7* (Fall): 2–10.
Cervero, Robert, and John Landis. 1997. Twenty Years of the Bay Area Rapid Transit System: Land Use and Development Impacts. *Transportation Research A: Policy and Practice 31,* 4: 309–333.
Cervero, Robert, and Carolyn Radisch. 1995. Travel Choices in Pedestrian Versus Automobile Oriented Neighborhoods. Working Paper 281. Berkeley: University of California Transportation Center (July).
Cervero, Robert, and Samuel Seskin. 1995. An Evaluation of the Relationships Between Transit and Urban Form. *Research Results Digest 7.* Washington, D.C.: Transit Cooperative Research Program, Transportation Research Board, National Research Council.
Cervero, Robert, and Kang-Li Wu. 1997. Polycentrism, Commuting, and Residential Location in the San Francisco Bay Area. *Environment and Planning A, 29:* 865–886.
Cervero, Robert, and Kang-Li Wu. 1998. Sub-Centring and Commuting: Evidence from the San Francisco Bay Area, 1980–1990. *Urban Studies 35,* 7: 1059–1076.
City of La Mesa. 1985a. *Grossmont Specific Plan.* La Mesa, Calif.: Community Development Department.
City of La Mesa. 1985b. *Report and Proposal to MTDB for Coordination of Development Along Amaya Drive in La Mesa.* Document 39202. La Mesa, Calif.: Office of City Clerk.
City of La Mesa. 1994. *Grossmont Specific Plan.* La Mesa, Calif.: Community Development Department.
City of Los Angeles. 1993. *Proposed Land Use/Transportation Policy.* Los Angeles: City Planning Department.
City of San Bernardino. 1993. *Land Use, Transportation and Air Quality: A Manual for Planning Practitioners.* Prepared by The Planning Center, Newport Beach, Calif.
City of San Diego. 1992. *Transit-Oriented Development Guidelines.* Prepared by Calthorpe Associates. San Diego: Development Services Department (October).
City of San Diego. 1993. *Rio Vista West Amendment to the First San Diego River Improvement Project Specific Plan Design Guidelines and Development Standards.* San Diego: Development Services Department (December 7).
City of San Diego. 1994. *First San Diego River Improvement Project Specific Plan, Sixth Amendment.* San Diego: Development Services Department (February 8).
Compin, Nicholas Shawn. 1996. *Rail Transit Station Development and the Municipal Land-Use Decision-Making Process.* M.A. Thesis. Irvine: University of California, Irvine School of Social Ecology.
Cone, Marla. 1996. Southland Smog Drops to Lowest Level in Decades. *Los Angeles Times* (Orange County Edition), October 30: A-1, A-12.
Consumer Reports. 1996. Neighborhoods Reborn. *61* 5.
Crane, Randall. 1996a. Cars and Drivers in the New Suburbs: Linking Access to Travel in Neotraditional Planning. *Journal of the American Planning Association 62,* 51–65.
Crane, Randall. 1996b. On Form Versus Function: Will the New Urbanism Reduce Traffic, or Increase It? *Journal of Planning Education and Research 15,* 117–126.
Crane, Randall. 1996c. The Influence of Uncertain Job Location on Urban Form and the Journey to Work. *Journal of Urban Economics 39,* 342–356.

Crane, Randall. 1998a. Travel by Design? *Access 12*, 2–7.
Crane, Randall. 1998b. *Suburbanization and Its Discontents: A Research Agenda.* Paper presented at Drachman Institute Conference on Urban and Suburban Growth, Phoenix (June).
Crane, Randall. 1999a. *Measuring the Impacts of Urban Form on Travel: Pitfalls and Promises.* Working Paper WP99RC1. Cambridge, Mass.: Lincoln Institute for Land Policy (January).
Crane, Randall. 1999b. *Poverty and Housing Consumption in the 1980s and 1990s: A Panel Study of the American Housing Survey.* Working Paper. Irvine: University of California (March).
Crane, Randall. 2000. The Influence of Urban Form on Travel: An Interpretive Review, *Journal of Planning Literature.*
Crane, Randall, and Richard Crepeau. 1998. Does Neighborhood Design Influence Travel? A Behavioral Analysis of Travel Diary and GIS Data. *Transportation Research D: Transport and Environment 3*, 225–238.
Crane, Randall, and Drusilla van Hengel. 1998. *Supply Versus Demand in Travel Access.* Working Paper. Berkeley: University of California Transportation Center.
Crane, Randall, and Roy Green. 1989. Debt Finance at the Municipal Level: Decision Making During the 1980s. *The Municipal Yearbook 1989.* Washington, D.C.: International City Management Association, 97–106.
Crane, Randall, and Amrita Daniere. 1996. Measuring Access to Basic Services in Global Cities: Descriptive and Behavioral Approaches. *Journal of the American Planning Association 62*, 203–221.
Cropper, Maureen L., and Wallace E. Oates. 1992. Environmental Economics: A Survey. *Journal of Economic Literature, 30*, 675–740.
Culot, Maurice, and Leon Krier. 1978. The Only Path for Architecture. *Oppositions 14*, 38–53.
Dahir, James. 1947. *The Neighborhood Unit Plan: Its Spread and Acceptance.* New York: Russell Sage Foundation.
Danielson, Michael N. 1976. *The Politics of Exclusion.* New York: Columbia University Press.
Danielson, Michael N., Alan M. Hershey, and John M. Bayne. 1977. *One Nation, So Many Governments.* Lexington, Mass.: Lexington Books.
Davis, Judy S., and Samuel Seskin. 1997. Impacts of Urban Form on Travel Behavior. *The Urban Lawyer 29*, 2: 215–232.
Deakin, Elizabeth. 1991. *Land Use and Transportation Planning in Response to Congestion: The California Experience.* Working Paper 54. Berkeley: University of California Transportation Center.
Deakin, Elizabeth, and Tilly Chang. 1992. *Barriers to Residential Development at Rail Transit Stations.* Berkeley: University of California Department of City and Regional Planning (mimeo).
Deka, Devajyoti, and Genevieve Giuliano. 1998. *What Can You Really Achieve with Community Design Standards?* Paper presented at Drachman Institute Conference on Urban and Suburban Growth, Phoenix (June).
Demoro, Harre W., and John N. Harder. 1989. *Light Rail Transit on the West Coast.* New York: Quadrant Press.
Dill, Jennifer. 1998. *Mandatory Employer-Based Trip Reduction: What Happened?* Paper presented at Transportation Research Board 77th annual meeting, Washington, D.C. (January).
DiMento, Joseph F., Sherry Ryan and Drusilla van Hengel. 1997. Local Government Land Use Policy Responses to the Century Freeway/Transitway, *Journal of Planning Education and Research 17*: 145–157.

Domencich, Thomas A., and Daniel McFadden. 1975. *Urban Travel Demand: A Behavioral Analysis.* Amsterdam: North Holland.
Downs, Anthony. 1962. The Law of Peak-Hour Expressway Congestion. *Traffic Quarterly 16,* 393–409.
Downs, Anthony. 1992. *Stuck in Traffic: Coping with Peak-Hour Traffic Congestion.* Washington, D.C.: The Brookings Institution.
Downs, Anthony. 1994. *New Visions for Metropolitan America.* Washington, D.C.: Brookings Institution, and Cambridge, Mass.: Lincoln Institute of Land Policy.
Dresch, Marla, and Steven M. Sheffrin. 1997. *Who Pays for Development Fees and Exactions?* San Francisco: Public Policy Institute of California.
Drummond, William J. 1995. Address matching: GIS technology for mapping human activity patterns. *Journal of the American Planning Association 61*: 240–251.
Duany, Andres. 1989. Traditional Towns. *Architectural Design 59*: 60–64.
Duany, Andres, and Elizabeth Plater-Zyberk. 1991. *Towns and Town-Making Principles.* New York: Rizzoli.
Duany, Andres, and Elizabeth Plater-Zyberk. 1992. The Second Coming of the American Small Town. *Wilson Quarterly* (Winter): 19–50.
Dunlop, Beth. 1989. Coming of Age. *Architectural Record 177* (July): 96–103.
Dunlop, Beth. 1991. Our Towns. *Architectural Record 179* (October): 110–119.
Dunphy, Robert T., and Kimberly Fisher. 1996. Transportation, Congestion, and Density: New Insights. *Transportation Research Record 1552,* 89–96.
Dunphy, Robert T., with Deborah L. Brett, Sandra Rosenbloom, and Andre Bald. 1997. *Moving Beyond Gridlock: Traffic and Development.* Washington, D.C.: The Urban Land Institute.
Ellwood, David. 1986. The Spatial Mismatch Hypothesis: Are There Teenage Jobs Missing in the Ghetto? Richard B. Freeman and Harry J. Holzer, eds., *The Black Youth Employment Crisis.* Chicago: University of Chicago Press.
Energy Information Administration. 1997. *Annual Energy Report.* Washington, D.C.: United States Department of Energy, Energy Information Administration (July).
Englin, J., and J. S. Shonkwiler. 1995. Modeling Recreation Demand in the Presence of Unobservable Travel Cots: Toward a Travel Price Model. *Journal of Environmental Economics and Management 29,* 3: 368–377.
Erbes, Russell E. 1996. *A Practical Guide to Air Quality Compliance,* 2nd ed. New York: Wiley.
Eskeland, Gunnar S., and Tarhan Feyzioglu. 1997. Rationing Can Backfire: The "Day Without a Car" in Mexico City. *The World Bank Economic Review 11,* 3: 383–408.
Ewing, Reid. 1994. Characteristics, Causes, and Effects of Sprawl: A Literature Review. *Environmental and Urban Issues 21,* 2: 1–15.
Ewing, Reid. 1995a. Measuring Transportation Performance. *Transportation Quarterly 49,* 1: 91–104.
Ewing, Reid. 1995b. Beyond Density, Mode Choice, and Single-Purpose Trips. *Transportation Quarterly 49,* 4: 15–24.
Ewing, Reid. 1997a. Is Los Angeles-Style Sprawl Desirable? *Journal of the American Planning Association 63.*
Ewing, Reid. 1997b. *Transportation & Land Use Innovations: When You Can't Pave Your Way Out of Congestion.* Chicago: Planners Press, American Planning Association.

Ewing, Reid, MaryBeth DeAnna, and Shi-Chiang Li. 1998. Land Use Impacts on Trip Generation Rates. *Transportation Research Record 1518,* 1–6.

Ewing, Reid, Padma Haliyur, and G. William Page. 1995. Getting Around a Traditional City, a Suburban Planned Unit Development, and Everything in Between. *Transportation Research Record 1466,* 53–62.

Fernandez, John M. 1994. Boulder Brings Back the Neighborhood Street. *Planning 60* (June): 21–26.

Fink, Marc. 1993. Toward a Sunbelt Urban Design Manifesto. *Journal of the American Planning Association 59*, 3: 320–333.

Fischel, William A. 1985. *The Economics of Zoning Laws.* Baltimore: Johns Hopkins Press.

Fischel, William A. 1989. Did Serrano Cause Proposition 13? *National Tax Journal 42,* 465–474.

Foster, Mark S. 1981. *From Streetcar to Superhighway: American City Planning and Urban Transportation, 1900–1940.* Philadelphia: Temple University Press.

Frank, Lawrence D., and Gary Pivo. 1995. Impacts of Mixed Use and Density on Utilization of Three Modes of Travel: Single-Occupant Vehicle, Transit, and Walking. *Transportation Research Record 1466,* 44–52.

Freeman, Ernest (Planning Director, San Diego. 1995. Personal interview. San Diego, Calif. (September 7).

Friedman, Bruce, Stephen P. Gordon, and John B. Peers. 1992. The Effect of Neotraditional Design on Travel Characteristics. *Compendium of Technical Papers.* Anchorage, Alaska: Institute of Transportation Engineers, 1992 District 6 Annual Meeting, 195–208.

Fritz, Niall (Director of Development Services, Santee, Calif.). 1995. Personal interview. Santee, Calif. (August 30).

Fujita, Masahisa. 1989. *Urban Economic Theory: Land Use and City Size.* Cambridge: Cambridge University Press.

Fulton, William. 1991. *Guide to California Planning.* Point Arena, Calif.: Solano Press Books.

Fulton, William. 1996. *The New Urbanism.* Cambridge, Mass.: Lincoln Institute of Land Policy.

Gärling, Tommy, Thomas Laitila, and Kerstin Westin, eds. 1998. *Theoretical Foundations of Travel Choice Modeling.* Amsterdam: Elsevier.

Garreau, Joel. 1991. *Edge City: Life on the New Frontier.* New York: Doubleday.

Garrett, Mark, and Martin Wachs. 1996. *Transportation Planning on Trial: The Clean Air Act and Travel Forecasting.* Thousand Oaks, Calif.: Sage.

Getter, Lisa. 1999. Cancer Risk from Air Pollution Still High, Study Says. *Los Angeles Times* (Orange County Edition), March 1: A-1, A-19.

Gibbs, W. Wayt. 1997. Transportation's Perennial Problems. *Scientific American 277* (October): 54–57.

Giuliano, Genevieve. 1989. New Directions for Understanding Transportation and Land Use. *Environment and Planning A, 21:* 145–159.

Giuliano, Genevieve. 1991. Is Jobs-Housing Balance a Transportation Issue? *Transportation Research Record 1305,* 305–312.

Giuliano, Genevieve. 1992. An Assessment of the Political Acceptability of Congestion Pricing. *Transportation 19,* 4: 335–358.

Giuliano, Genevieve. 1995a. Land Use Impacts of Transportation Investments: Highway and Transit. Susan Hanson, ed., *The Geography of Travel,* 2nd ed. New York: Guilford Press, 305–341.

Giuliano, Genevieve. 1995b. The Weakening Transportation-Land Use Connection. *Access 6* (Spring): 3–11.

Giuliano, Genevieve, Keith Hwang, and Martin Wachs. 1993. Employee Trip Reduction in Southern California: First Year Results. *Transportation Research 27A,* 2: 125–137.

Giuliano, Genevieve, D. W. Levine, and R. F. Teal. 1990. Impact of High Occupancy Vehicle Lanes on Carpooling Behavior. *Transportation 17,* 159–177.

Giuliano, Genevieve, and Kenneth A. Small. 1991. Subcenters in the Los Angeles Region. *Regional Science and Urban Economics 21,* 163–182.

Giuliano, Genevieve, and Kenneth A. Small. 1993. Is the Journey to Work Explained by Urban Structure? *Urban Studies 30,* 9: 1485–1500.

Glazer, Amihai, Daniel B. Klein, and Charles Lave. 1995. Clean on Paper, Dirty on the Road: Troubles with California's Smog Check. *Journal of Transport Economics and Policy 29,* 85–92.

Gold, Steven D., and Judy A. Zelio. 1990. *State-Local Fiscal Indicators.* Washington, D.C.: National Conference of State Legislators.

Gómez-Ibáñez, José A. 1985. A Dark Side to Light Rail? The Experience of Three New Transit Systems. *Journal of the American Planning Association 51,* 337–351.

Gómez-Ibáñez, José A., and John R. Meyer. 1993. *Going Private: The International Experience with Transport Privatization.* Washington, D.C.: The Brookings Institution.

Gómez-Ibáñez, José A., and Kenneth A. Small. 1994. *Road Pricing for Congestion Management: A Survey of International Practice.* National Cooperative Highway Research Program, Synthesis of Highway Practice 210. Washington, D.C.: National Academy Press.

Gordon, Peter, Ajay Kumar, and Harry W. Richardson. 1989a. The Influence of Metropolitan Spatial Structure on Commuting Time. *Journal of Urban Economics 26,* 138–151.

Gordon, Peter, Ajay Kumar, and Harry W. Richardson. 1989b. Gender Differences in Metropolitan Travel Behavior. *Regional Studies 23,* 499–510.

Gordon, Peter, Ajay Kumar, and Harry W. Richardson. 1989c. The Spatial Mismatch Hypothesis—Some New Evidence. *Urban Studies 26,* 315–326.

Gordon, Peter, and Harry W. Richardson. 1993. The Facts About "Gridlock" in Southern California. Policy Study 165. Los Angeles: Reason Foundation.

Gordon, Peter, and Harry W. Richardson. 1996. Employment Decentralization in U.S. Metropolitan Areas: Is Los Angeles an Outlier or Norm? *Environment and Planning A, 28:* 1727–1743.

Gordon, Peter, Harry W. Richardson, and Myung-Jin Jun. 1991. The Commuting Paradox: Evidence from the Top Twenty. *Journal of the American Planning Association 57,* 4: 416–420.

Gordon, Peter, Harry W. Richardson, and Yu-Chun Liao. 1997. A Note on the Travel Speeds Debate. *Transportation Research 31A,* 3: 259–262.

Gramlich, Edward M. 1994. Infrastructure Investment: A Review Essay. *Journal of Economic Literature 32,* 3: 1176–1196.

Greene, William. 1993. *Econometric Analysis,* 2nd ed. New York: Macmillan.

Griffin, James (Director of Community Development, El Cajon, Calif.). 1995. Personal interview. El Cajon, Calif. (August 28).

Gyourko, Joseph, and Richard Voith. 1999. The Tax Treatment of Housing and Its Effects on Bounded and Unbounded Communities. Working Paper WP99JG2. Cambridge, Mass.: Lincoln Institute of Land Policy.

Haab, Timothy C., and Robert L. Hicks. 1997. Accounting for Choice Set Endogeneity in Random Utility Models of Recreation Demand. *Journal of Environmental Economics and Management 34,* 127–147.

Hahn, Kenneth. 1967. *A Factual Record of Correspondence Between Kenneth Hahn, Los Angeles County Supervisor, and the Presidents of General Motors, Ford and Chrysler Regarding the Automobile Industry's Obligation to Meet Its Rightful Responsibility in Controlling Air Pollution from Automobiles.* Los Angeles: County of Los Angeles.

Handy, Susan. 1992. Regional Versus Local Accessibility: Neo-Traditional Development and Its Implications for Nonwork Travel. *Built Environment 18,* 253–267.

Handy, Susan. 1993. Regional Versus Local Accessibility: Implications for Nonwork Travel. *Transportation Research Record,* No. 1400: 58–66.

Handy, Susan. 1996a. Understanding the Link Between Urban Form and Nonwork Travel Behavior. *Journal of Planning Education and Research 15,* 3: 183–198.

Handy, Susan. 1996b. Methodologies for Exploring the Link Between Urban Form and Travel Behavior. *Transportation Research D: Transport and Environment 1,* 2: 151–165.

Handy, Susan. 1997. *Travel Behavior—Land Use Interactions: An Overview and Assessment of the Research.* Working paper. Austin: University of Texas (March).

Handy, Susan, Kelly Clifton, and Janice Fisher. 1998. *The Effectiveness of Land Use Policies as a Strategy for Reducing Automobile Dependence: A Study of Austin Neighborhoods.* Research Report SWUTC/98/467501–1. University of Texas (October).

Handy, Susan, and Deborah A. Niemeier. 1997. Measuring Accessibility: An Exploration of Issues and Alternatives. *Environment and Planning A, 29:* 1175–1194.

Hansen, Mark, and Yuanlin Huang. 1997. Road Supply and Traffic in California Urban Areas. *Transportation Research 31A,* 3: 205–218.

Hansen, W. G. 1959 How Accessibility Shapes Land Use. *Journal of the American Institute of Planners 25,* 73–76.

Hanson, Susan, and Margo Schwab. 1987. Accessibility and Intraurban Travel. *Environment and Planning A, 19:* 735–748.

Heikkila, Eric J. 1994. Microeconomics and Planning: Using Simple Diagrams to Illustrate the Economics of Traffic Congestion. *Journal of Planning Education and Research 14,* 1: 29–41.

Hise, Greg. 1997. *Magnetic Los Angeles: Planning the Twentieth-Century Metropolis.* Baltimore: Johns Hopkins University Press.

Holtzclaw, John. 1990. *Explaining Urban Density and Transit Impacts on Auto Use.* Working paper. Natural Resources Defense Council (April).

Holtzclaw, John. 1994. *Using Residential Patterns and Transit to Decrease Auto Dependence and Costs.* Working paper. San Francisco: Natural Resources Defense Council.

Howitt, Arnold M., and Alan Altshuler. 1999. The politics of controlling auto air pollution, in José Gómez-Ibáñez, William B. Tye and Clifford Winston, eds. *Essays in Transportation Economics and Policy: A Handbook in Honor of John R. Meyer.* Washington, D.C.: Brookings Institution: 223–255.

Inman, Bradley. 1993. Public Likes Neotraditional Style, *Los Angeles Times,* August 29: K9.

Institute of Transportation Engineers. 1991. *Trip Generation, ITE,* 5th ed. Washington, D.C.: ITE.

Institute of Transportation Engineers. 1997. *Traditional Neighborhood Development: Street Design Guidelines. A Proposed Recommended Practice of the ITE.* Washington, D.C.: ITE.

Jacobs, Jane. 1961. *The Death and Life of Great American Cities.* New York: Random House.

JHK & Associates. 1995. Transportation-Related Land Use. *Strategies to Minimize Motor Vehicle Emissions: An Indirect Source Research Study, Final Report.* Sacramento: California Air Resources Board (June).

Johnston, John, and John DiNardo. 1996. *Econometric Methods, Fourth Edition,* New York: McGraw-Hill, October 1996.

Johnston, Robert A., and Raju Ceerla. 1995. *Effects of Land Use Intensification and Auto Pricing Policies on Regional Travel, Emissions, and Fuel Use.* Working Paper 269. Berkeley: University of California Transportation Center.

Jones, Maryanne, and Mary Jane Breinholt. 1993. *The Role of Land Use Strategies for Improving Transportation and Air Quality: Summary of Proceedings.* Los Angeles: Public Policy Program, UCLA Extension.

Kain, John F. 1968. Housing Segregation, Negro Employment, and Metropolitan Decentralization. *Quarterly Journal of Economics 82,* 175–197.

Kain, John. 1999. The Urban Transportation Problem: A Reexamination and Update, in José Gómez-Ibáñez, William B. Tye and Clifford Winston, eds. *Essays in Transportation Economics and Policy: A Handbook in Honor of John R. Meyer.* Washington, D.C.: Brookings Institution: 359–401.

Kain, John F., and Gary R. Fauth. 1976. *The Effects of Urban Structure on Household Auto Ownership Decisions and Journey to Work Mode Choice.* Research Report R76-1. Cambridge, Mass.: Department of City and Regional Planning, Harvard University (May).

Kain, John F., and Gary R. Fauth. 1977. *The Impact of Urban Development on Auto Ownership and Transit Use.* Discussion Paper D77-18. Cambridge: Department of City and Regional Planning, Harvard University (December).

Kaplan, Sam H. 1990. The Holy Grid: A Skeptic's View. *Planning 56,* 10–11.

Kasarda, John. 1995. Industrial Restructuring and Changing Location of Jobs. R. Farley, ed., *State of the Union: America in the 1990s, Vol. 1: Economic Trends.* New York: Russell Sage Foundation, 215–266.

Katz, Peter. 1994. *The New Urbanism: Toward an Architecture of Community.* New York: McGraw-Hill.

Keeler, Theodore E., and Kenneth A. Small. 1977. Optimal Peak Load Pricing, Investment, and Service Levels on Urban Expressways. *Journal of Political Economy 85,* 1: 1–25.

Kelbaugh, Doug, ed. 1989. *The Pedestrian Pocket Book: A New Suburban Design Strategy.* New York: Princeton Architectural Press.

Kitamura, Ryuichi, and Wilfred W. Recker. 1985. Activity-Based Travel Analysis. In Gijsbertus, R. M. Jansen, Peter Nijkamp, and Cees J. Ruijgrok, eds., *Transportation and Mobility in an Era of Transition.* Amsterdam: Elsevier.

Kitamura, Ryuichi, Cynthia Chen, and Ram M. Pendyala. 1997. Generation of Synthetic Daily Activity-Travel Patterns. *Transportation Research Record 1607,* 154–162.

Kitamura, Ryuichi, Laura Laidet, Patricia Mokhtarian, Carol Buckinger, and Fred Gianelli. 1994. *Land Use and Travel Behavior.* Report UCD-ITS-RR-94–27. Davis: Institute of Transportation Studies, University of California (October).

Kitamura, Ryuichi, Patricia Mokhtarian, and Laura Laidet. 1997. A Micro-Analysis of Land Use and Travel in Five Neighborhoods in the San Francisco Bay Area. *Transportation 24,* 125–158.

Knack, Ruth Eckdish. 1989. Repent, Ye Sinners, Repent: Neotraditional

Town Planning in the Suburbs. *Planning 55* (August): 4–13.
Knack, Ruth E. 1995. BART's Village Vision. *Planning 61* (January): 18–21.
Knack, Ruth E. 1998. Drive Nicely. *Planning 64* (December): 12–15.
Knight, Robert, and Lisa Trygg. 1977. Evidence of Land Use Impacts of Rapid Transit Systems. *Transportation 6,* 231–247.
Kockelman, Kara Maria. 1997. Travel Behavior as Function of Accessibility, Land Use Mixing, and Land Use Balance: Evidence from San Francisco Bay Area. *Transportation Research Record 1607,* 116–125.
Kockelman, Kara Maria. 1998. *A Utility-Theory-Consistent System-of-Demand-Equations Approach to Household Travel Choice.* Unpublished Ph.D. Dissertation. Berkeley: Department of Civil and Environmental Engineering, University of California.
Koppelman, Frank S., Chandra R. Bhat, and Joseph L. Schofer. 1993. Market Research Evaluation of Actions to Reduce Suburban Traffic Congestion: Commuter Travel Behavior and Response to Demand Reduction Actions. *Transportation Research 27A,* 5: 383–393.
Kostof, Spiro. 1991. *The City Shaped: Urban Patterns and Meanings Through History.* London: Thames and Hudson.
Kostof, Spiro. 1992. *The City Assembled: The Elements of Urban Form Through History.* Boston: Bullfinch Press.
Kreps, D. M. 1990. *A Course in Microeconomic Theory.* Princeton: Princeton University Press.
Krugman, Paul. 1991. *Geography and Trade.* Cambridge, Mass.: MIT Press.
Kulash, Walter, Joe Anglin, and David Marks. 1990. Traditional Neighborhood Development: Will the Traffic Work? *Development 21* (July/August): 21–24.
Kulkarni, Anup Arvind. 1996. *The Influence of Land Use and Network Structure on Travel Behavior.* Unpublished Master's Dissertation. Irvine: University of California, Department of Civil and Environmental Engineering.
Kunstler, James Howard. 1993. *The Geography of Nowhere: The Rise and Decline of America's Man-Made Landscape.* New York: Simon & Schuster.
Ladd, Helen F. 1975. Local education expenditures, fiscal capacity and the composition of the property tax base. *National Tax Journal 28:* 145–158.
Ladd, Helen F., and John Yinger. 1989. *America's Ailing Cities: Fiscal Health and the Design of Urban Policy.* Baltimore: Johns Hopkins University Press.
Landis, John, and Robert Cervero. 1999. Middle Age Sprawl: BART and Urban Development. *Access 14* (Spring): 2–15.
Lave, Lester B., and Gilbert S. Omenn. 1981. *Clearing the Air: Reforming the Clean Air Act.* Washington, D.C.: The Brookings Institution.
Leccese, Michael. 1990. Next Stop: Transit-Friendly Towns. *Landscape Architecture 80*: 47–53.
Leiter, Robert (Planning Director, Chula Vista, California). 1995. Personal interview. Chula Vista, Calif. (August 30).
Lerner-Lam, Eva, Stephen P. Celniker, Gary W. Halbert, Chester Chellman, and Sherry Ryan. 1991. Neo-Traditional Neighborhood Design and Its Implications for Traffic Engineering. *Institute of Transportation Engineers 1991 Compendium of Technical Papers.* Washington, D.C.: ITE, 135–141.
Levine, Jonathan C. 1998. Rethinking Accessibility and Jobs-Housing Balance. *Journal of the American Planning Association 64,* 2: 133–149.
Levinson, Arik, and Sudhir Shetty. 1992. *Efficient Environmental Regulation: Case Studies of Urban Air Pollution in Los Angeles, Mexico City, Cubatao, and Ankara.* World Bank Policy Research Working Paper Series. Washington, D.C.: The World Bank.

Levinson, David M., and Ajay Kumar. 1994. The Rational Locator: Why Travel Times Have Remained Stable. *Journal of the American Planning Association 60,* 3: 319–332.

Levinson, David, and Ajay Kumar. 1995. Activity, Travel, and the Allocation of Time. *Journal of the American Planning Association 61,* 4: 458–470.

Levinson, David, and Ajay Kumar. 1997. Density and the Journey to Work. *Growth and Change 28* (Spring): 147–172.

Lewis, Paul, and Elisa Barbour. 1999. *California Cities and the Local Sales Tax.* San Francisco: Public Policy Institute of California.

Lynch, Kevin. 1981. *A Theory of Good City Form.* Cambridge: MIT Press.

Lindblom, Charles E. 1959. The Science of "Muddling Through." *Public Administration Review 19,* 79–88.

Lindley, Jeffrey A. 1987. Urban Freeway Congestion: Quantification of the Problem and Effectiveness of Potential Solutions. *ITE Journal* (January): 27–31.

Linneman, Peter, and Phillip E. Graves. 1983. Migration and Job Change: A Multinomial Logit Approach. *Journal of Urban Economics 14,* 263–279.

Los Angeles Times (Orange County edition). 1992. Smog Season Ends with 41 Stage I Alerts—A Low Total. October 31: B-3.

Losch, August. 1954. *The Economics of Location,* trans. by W. Woglum and W. Stolper. New Haven, Conn.: Yale University Press.

Madden, Janice, and C. Lic (1990) The Wage Effects of Residential Location and Commuting Constraints on Employed Married Women. *Urban Studies 27,* 353–369.

Madden, Janice, and Michelle White. 1980. Spatial Implications of Increases in the Female Labor Force. *Land Economics 56,* 432–446.

Mastako, Kimberly Allyn, Laurence R. Rilett, and Edward C. Sullivan. 1998. *Commuter Behavior on California State Route 91—After Introducing Variable-Toll Express Lanes.* Paper presented at Transportation Research Board 77th annual meeting, Washington, D.C. (January 11–15).

McFadden, Daniel. 1996. *On Computation of Willingness-to-Pay in Travel Demand Models.* Working paper. Berkeley: Department of Economics, University of California (July).

McNally, Michael G. 1993. *Regional Impacts of Neotraditional Neighborhood Development.* Working Paper No. 172. Berkeley: University of California Transportation Center (May).

McNally, Michael G. 1997. An Activity-Based Microsimulation Model for Travel Demand Forecasting. In Dick Ettema and Harry Timmermans, eds., *Activity-Based Approaches to Travel Analysis.* Oxford: Pergamon.

McNally, Michael G., and Anup Kulkarni. 1997. Assessment of Influence of Land Use-Transportation System on Travel Behavior. *Transportation Research Record 1607,* 105–115.

McNally, Michael G., and Sherry Ryan. 1993. Comparative Assessment of Travel Characteristics for Neotraditional Designs. *Transportation Research Record 1400,* 67–77.

Messenger, Todd, and Reid Ewing. 1996. Transit-Oriented Development in the Sunbelt. *Transportation Research Record 1552.*

Metropolitan Transit Development Board. N.d.-a. *MTDB Joint Development Project Sites.* San Diego: Metropolitan Transit Development Board.

Metropolitan Transit Development Board. N.d.-b. *Transit Linking San Diego's Development.* San Diego: Metropolitan Transit Development Board.

Meyer, John R., and José A. Gómez-Ibáñez. 1981. *Autos, Transit, and Cities.* Cambridge, Mass.: Harvard University Press.

Meyer, John R., John Kain, and Martin Wohl. 1965. *The Urban Transportation Problem.* Cambridge, Mass.: Harvard University Press.
Mieszkowski, Peter, and Mills, Edwin S. 1993. The Causes of Metropolitan Suburbanization. *Journal of Economic Perspectives 7,* 3: 135–147.
Mills, Edwin S., 1972. *Urban Economics.* Glenview, Ill.: Scott, Foresman.
Mills, Edwin S., and Wallace E. Oates, eds. 1975. *Fiscal Zoning and Land Use Controls: The Economic Issues.* Lexington, Mass.: Lexington Books.
Mitchell, Robert B., and Chester Rapkin. 1954. *Urban Traffic: A Function of Land Use.* New York: Columbia University Press.
Mohney, David, and Keller Easterling, eds. 1991. *Seaside: Making a Town in America.* New York: Princeton Architectural Press.
Mohring, Herbert, and Mitchell Harwitz. 1962. *Highway Benefits: An Analytical Framework.* Evanston, Ill.: Northwestern University Press.
Mokhtarian, Patricia L., Elizabeth A. Raney, and Ilan Salomon. 1997. Behavioral Responses to Congestion: Identifying Patterns and Socio-Economic Differences in Adoption. *Transport Policy 7,* 3: 143–160.
Moon, Henry. 1990. Land Use Around Suburban Transit Stations. *Transportation 17,* 67–88.
Moore, Terry, and Paul Thorsnes. 1994. *The Transportation/Land Use Connection: A Framework for Practical Policy.* Chicago: American Planning Association, Planning Advisory Service.
Morey, Edward R., and Donald M. Waldman. 1998. Measurement Error in Recreation Demand Models: The Joint Estimation of Participation, Site Choice, and Site Characteristics. *Journal of Environmental Economics and Management 35,* 3: 262–276.
Murphy, James J., and Mark A. Delucchi. 1998. A Review of the Literature on the Social Cost of Motor Vehicle Use in the United States. *Journal of Transportation and Statistics* (January): 15–42.
Muth, Richard. 1969. *Cities and Housing.* Chicago: University of Chicago Press.
Myers, Dowell, and Alicia Kitsuse. 1999. *The Debate over Future Density of Development: An Interpretive Review.* Working Paper WP99DM1. Cambridge, Mass.: Lincoln Institute of Land Policy.
Nelson, Arthur C., ed. 1988. *Development Impact Fees: Policy Rationale, Practice, Theory, and Issues.* Chicago: Planners Press, American Planning Association.
Newman, Peter W. G., and Jeffrey R. Kenworthy. 1989. *Cities and Automobile Dependence: An International Sourcebook.* Brookfield, Vt.: Gower.
Noland, Robert B. 1997. Commuter Responses to Travel Time Uncertainty Under Congested Conditions: Expected Costs and the Provision of Information. *Journal of Urban Economics 41:* 377–406.
Noland, Robert B. 1998. Relationships between Highway Capacity and Induced Vehicle Travel. Working paper. Washington, D.C.: United States Environmental Protection Agency.
Nowland, D. M., and G. Stewart. 1991. Downtown Population Growth and Commuting Trips: Recent Experience in Toronto. *Journal of the American Planning Association 57,* 165–182.
NPTS. 1993. The 1990 Nationwide Personal Transportation Survey Databook, prepared by Oak Ridge National Laboratory for the Office of Highway Information Management. Washington, D.C.: U.S. Dept. of Transportation, Federal Highway Administration.
Olmsted, Frederick Law. 1924. *A Major Traffic Street Plan for Los Angeles.*

Report prepared for the Committee on Los Angeles Plan of Major Highways of the Traffic Commission of the City and County of Los Angeles.

Ong, Paul, and Evelyn Blumenberg (1998). Job Access, Commute, and Travel Burden Among Welfare Recipients. *Urban Studies 35*, 1: 77–94.

Orange County Transportation Authority. 1991. *Long Range Transit Systems Plan and Development Strategy.* Final Report of the Countywide Rail Study, submitted by Parsons Brinckerhoff. Los Angeles (October).

Orange County Transportation Commission (OCTC). 1980. *Orange County Multimodal Transportation Study Refinement Project.* Analysis and Evaluation of MMTS Selected System technical report. Los Angeles: OCTC (October).

O'Regan, Katherine M., and John M. Quigley. 1999. Accessibility and Economic Opportunity, in José Gómez-Ibáñez, William B. Tye and Clifford Winston, eds. *Essays in Transportation Economics and Policy: A Handbook in Honor of John R. Meyer.* Washington, D.C.: Brookings Institution: 437–466.

Ortner, James (Manager of Intergovernmental Relations and Air Quality, Orange County Transportation Authority) 1996. Telephone interview. (May 22).

Ortuzar, Juan de D., and Luis G. Willumsen. 1994. *Modeling Transport.* New York: Wiley.

O'Toole, Randal. 1996. Packing 'Em In: Metro Plan for Urban Area May Not Be What's Wanted—or Needed. *The Oregonian* (Portland), November 17.

Park, Robert Ezra. 1952. *Human Communities: The City and Human Ecology.* Glencoe, Ill.: Free Press.

Parrish, David. 1997. OCTA Votes to Continue Toll Study. *Orange County Register,* July 29.

Parsons, G. R., and M. J. Kealy. 1995. A Demand Theory for Number of Trips in a Random Utility Model of Recreation. *Journal of Environmental Economics and Management 29,* 3: 357–367.

Pendall, Rolf. 1999. Do Land Use Controls Cause Sprawl? *Environment and Planning B: Planning and Design.*

Peng, Zhong-Ren. 1997a. *Travelers' Stated Choices to Cope with Traffic Congestion: Nonwork Trips.* Working paper. Milwaukee: University of Wisconsin. (July).

Peng, Zhong-Ren. 1997b. *Travelers' Stated Choices to Cope with Traffic Congestion: Journey to Work Commute.* Working paper. Milwaukee: University of Wisconsin. (July).

Perry, Clarence. 1939. *Housing for the Machine Age.* New York: Russell Sage Foundation.

Pickrell, Don. 1999. Transportation and Land Use, in José Gómez-Ibáñez, William B. Tye and Clifford Winston, eds. *Essays in Transportation Economics and Policy: A Handbook in Honor of John R. Meyer.* Washington, D.C.: Brookings Institution: 403–435.

Pipkin, John S. 1995. Disaggregate Models of Travel Behavior. Susan Hanson, ed., *The Geography of Travel,* 2nd ed. New York: Guilford Press, 188–218.

Pivo, Gary, Paul Hess, and Abhay Thatte. 1995. *Land Use Trends Affecting Auto Dependence in Washington's Metropolitan Areas, 1970–1990.* Report WA-RD 380.1. Olympia: Washington State Department of Transportation (July).

Pivo, Gary, Anne V. Moudon, Paul Hess, Kit Perkins, Lawrence Frank, Franz E. Loewenherz, and Scott Rutherford. 1992. *A Strategic Plan for Research-*

ing Urban Form Impacts on Travel Behavior. Report WA-RD 261.1. Olympia: Washington State Department of Transportation (December).

Pivo, Gary, Anne V. Moudon, and Franz E. Loewenherz. 1992. *A Summary of Guidelines for Coordinating Urban Design, Transportation and Land Use Planning, with an Emphasis on Encouraging Alternatives to Driving Alone.* Report WA-RD 261.3. Olympia: Washington State Department of Transportation (August).

Plane, David. 1995. Urban Transportation: Policy Alternatives. Susan Hanson, ed., *The Geography of Travel,* 2nd ed. New York: Guilford Press, 435–469.

Post, Roger (Planning Director, National City, California). 1995. Personal interview. National City, California, August 29.

Pund, Ernest, E. 1997. Way Down South. *Riverside Press Enterprise,* March 10: 13.

Pushkarev, B., and J. Zupan. 1977. *Public Transportation and Land Use Policy.* Bloomington: Indiana University Press.

Quigley, John M., and Daniel H. Weinberg. 1977. Intra-Urban Residential Mobility: A Review and Synthesis. *International Regional Science Review 2,* 41–66.

Rabiega, W. A., and D. A. Howe. 1994. Shopping Travel Efficiency of Traditional, Neotraditional, and Cul-de-Sac Neighborhoods with Clustered and Strip Commercial, Paper presented at the Association of Collegiate Schools of Planning 36th Annual Conference, Tempe.

Reckhard, E. Scott. 1998. Back at Drawing Boards to Build a Better Yesterday, *Los Angeles Times,* April 26, p. 3.

Roberts, John, and Chris Wood. 1992. *Land Use and Travel Demand.* Proceedings of Seminar B, PTRC Transport, Highways and Planning, summer annual meeting. Manchester: University of Manchester (September).

Rose, Mark H. 1990. *Interstate: Express highway Politics, 1939–1989.* Knoxville: University of Tennessee Press.

Rosenbloom, Sandra. 1993. *Travel by Women.* Tucson: University of Arizona, Drachman Institute for Land and Regional Development Studies.

Rothenberg, Jerome. 1970. The Economics of Congestion and Pollution: An Integrated View, *American Economic Review 60,* Papers and Proceedings, 114–121.

Rowe, Peter G. 1991. *Making a Middle Landscape.* Cambridge, Mass.: MIT Press.

Rutherford, G. Scott, Edward McCormack, and Martina Wilkinson. 1996. *Travel Impacts of Urban Form: Implications from an Analysis of Two Seattle Area Travel Diaries.* Paper prepared for the Travel Model Improvement Program Conference on Urban Design, Telecommuting and Travel Behavior, University of Washington, (October). Seattle

Ryan, Sherry, and Michael G. McNally. 1995. Accessibility of Neotraditional Neighborhoods: A Review of Design Concepts, Policies, and Recent Literature. *Transportation Research A, 29*: 87–105.

Samuelson, Paul. 1954. The Pure Theory of Public Expenditures. *Review of Economics and Statistics 36,* 4: 386–389.

SANDAG. 1987a. Sourcepoint: 1986 Land Use Inventory, January, San Diego, CA: San Diego Association of Governments.

SANDAG. 1987b. 1986 Travel Behavior Surveys, Volume I; Results of Surveys, September, San Diego, CA: San Diego Association of Governments.

SANDAG. 1987c. 1986 Travel Behavior Surveys, Volume II; Technical Documentation, September, San Diego, CA: San Diego Association of Governments.

Sarmiento, Sharon. 1995. *Studies in Transportation and Residential Mobility.* Unpublished Ph.D. Dissertation. University of California, Irvine, Department of Economics.
Schimek, Paul. 1998. Daily Travel as a Function of Population Density. Volpe Center, U.S. DOT, Cambridge.
Schneider, Mark. 1989, *The Competitive City: The Political Economy of Suburbia.* Pittsburgh: University of Pittsburgh Press.
Schrank, David L., and Timothy J. Lomax. 1997. *Urban Roadway Congestion in Major Urban Areas, 1982–1994.* Research Report 1131–9. College Station, Tex.: Texas Transportation Institute.
Shen, Qing. 1998. *Uncovering Spatial and Social Dimensions of Commuting.* Working Paper. Cambridge, Mass.: MIT (August).
Shoup, Donald C. 1997. Evaluating the Effects of Cashing Out Employer-Paid Parking: Eight Case Studies. *Transport Policy 4,* 4: 201–216.
Shoup, Donald C. 1999. In Lieu of Required Parking. *Journal of Planning Education and Research 18,* 4: 307–320.
Shrouds, James M. 1992. Transportation Planning Requirements of the Federal Clean Air Act: A Highway Perspective. Roger L. Wayson, ed., *Transportation Planning and Air Quality.* New York: American Society of Civil Engineers.
Small, Kenneth A. 1992. *Urban Transportation Economics.* Chur, Switzerland: Harwood.
Small, Kenneth A., and José A. Gómez-Ibáñez. 1996. *Urban Transportation.* Working Paper 97–3. Irvine, California: University of California, Irvine Institute of Transportation Studies.
Small, Kenneth A., and Cheng Hsiao. 1985. Multinomial Logit Specification Tests. *International Economic Review 26,* 3: 619–627.
Small, Kenneth A., and Camilla Kazimi. 1995. On the Costs of Air Pollution from Motor Vehicles. *Journal of Transport Economics and Policy 29,* 1: 7–32.
Small, Kenneth A., and Clifford Winston. 1999. The demand for transportation: Models and applications, in José Gómez-Ibáñez, William B. Tye and Clifford Winston, eds. *Essays in Transportation Economics and Policy: A Handbook in Honor of John R. Meyer.* Washington, D.C.: Brookings Institution: 11–55.
Small, Kenneth A., Clifford Winston, and Carol A. Evans. 1989. *Road Work: A New Highway Pricing and Investment Policy.* Washington, D.C.: The Brookings Institution.
Solow, Robert M. 1973. On Equilibrium Models of Urban Location. J. M. Parkins, ed., *Essays in Modern Economics.* London: Longman.
South Coast Air Quality Management District. 1997. *The Southland's War on Smog: Fifty Years of Progress Toward Cleaner Air.* Diamond Bar, Calif.: South Coast Air Quality Management District.
Southworth, Michael. 1997. Walkable Suburbs? An Evaluation of Neotraditional Communities at the Urban Edge. *Journal of the American Planning Association 63,* 1: 28–44.
Southworth, Michael, and Eran Ben-Joseph. 1995. Streets and the Shaping of Towns and Cities. *Journal of the American Planning Association 61,* 1: 65–81.
Southworth, Michael, and Eran Ben-Joseph. 1997. *Streets and the Shaping of Towns and Cities.* New York: McGraw-Hill.
Southworth, Michael, and Peter M. Owens. 1993. The Evolving Metropolis: Studies of Community, Neighbor-hood, and Street Form at the Urban Edge. *Journal of the American Planning Association 59,* 3: 271–287.

Spillar, Robert J., and G. Scott Rutherford. 1990. The Effects of Population Density and Income on Per Capita Transit Ridership in Western American Cities. *Compendium of Technical Papers.* Boise, Idaho: Institute of Transportation Engineers.

Steiner, Ruth L. 1994. Residential Density and Travel Patterns: Review of the Literature. *Transportation Research Record 1466,* 47–43.

Steiner, Ruth L. 1995. *Mode Choice for Shopping in Traditional Neighborhoods: Factors Influencing the Decision to Walk.* Working Paper. Gainesville: University of Florida.

Stone, J. R., M. D. Foster, and C. E. Johnson. 1992. Neo-Traditional Neighborhoods: A Solution to Traffic Congestion? *Site Impact Traffic Assessment.* New York: American Society of Civil Engineers.

Stone, Keith. 1996. Breaking Driving Habits. *Los Angeles Daily News,* August 4: 1.

Sullivan, Edward C., and Joe El Harake. 1998. *The California Route 91 Toll Lanes—Observed Impacts and Other Observations.* Paper presented at Transportation Research Board 77th annual meeting, Washington, D.C. (January).

Sun, Xiaoduan, Chester G. Wilmot, and T. Kasturi. 1998. *Household Travel, Household Characteristics, and Land Use: An Empirical Study from the 1994 Portland Travel Survey.* Working Paper: University of Southwestern Louisiana and Louisiana State University.

Taylor, Brian D. 1995. Public Perceptions, Fiscal Realities, and Freeway Planning: The California Case. *Journal of the American Planning Association 61,* 1: 43–56.

Taylor, Brian D., and Paul M. Ong. 1995. Spatial Mismatch or Automobile Mismatch? An Examination of Race, Residence and Commuting in US Metropolitan Areas. *Urban Studies 32,* 9: 1453–1473.

Tertoolen, Gerard, Dik van Kreveld, and Ben Verstraten. 1998. Psychological Resistance Against Attempts to Reduce Private Car Use. *Transportation Research A 32,* 3: 171–181.

Thomas Guide Street Guide and Directory. 1994.

Thompson, Gregory L., and James E. Frank. 1995. *Evaluating Land Use Methods for Altering Travel Behavior.* Report No. NUTI93FSU4.1–2. Tallahassee: National Urban Transit Institute, Florida State University (January).

Tiebout, Charles. 1956. A Pure Theory of Local Expenditures. *The Journal of Political Economy 64,* 5: 416–424.

Tirole, Jean. 1988. *The Theory of Industrial Organization.* Cambridge, Mass.: MIT Press.

Topp, H. 1993. Parking Policies to Reduce Car Traffic in German Cities. *Transport Reviews 13,* 1: 83–95.

Train, Kenneth. 1986. *Qualitative Choice Analysis: Theory, Econometrics, and an Application to Automobile Demand.* Cambridge, Mass.: MIT Press.

Train, Kenneth E. 1998. Recreation Demand Models with Taste Differences over People. *Land Economics 74,* 2: 230–239.

Trombley, William. 1986. San Diego on a Roll with New Trolley Line. *The Los Angeles Times,* March 21: 3.

Untermann, Richard K. 1984. *Accommodating the Pedestrian: Adapting towns and Neighborhoods for Walking and Bicycling.* New York: Van Nostrand Reinhold.

U.S. Department of Transportation. 1992. *Exploring Key Issues in Public-Private Partnerships for Highway Development.* Policy Discussion Paper 2,

Searching for Solutions Series. Washington, D.C.: U.S. Department of Transportation (June).
U.S. Department of Transportation, Bureau of Transportation Statistics. 1997. *Transportation Statistics Annual Report, 1997.* Washington, D.C.: U.S. GPO.
U.S. Environmental Protection Agency. 1996. *National Air Quality Emissions and Trends Report, 1995.* Research Triangle Park, N.C.: U.S. Environmental Protection Agency, Air Quality Trends Analysis Group.
U.S. Office of Management and Budget. 1997. *Analytical Perspectives: Budget of the United States Government, Fiscal Year 1998.* Washington, D.C.: U.S. GPO.
Van Ommeren, Jon, Peit Rietveld, and Peter Nijkamp. 1997. Commuting: In Search of Jobs and Residences. *Journal of Urban Economics 42,* 402–421.
Vickerman, Roger W. 1972. The Demand for Nonwork Travel. *Journal of Transport Economics and Policy.*
Vickrey, William S. 1963. Pricing in Urban and Suburban Transport. *American Economic Review 53,* 2: 452–465.
Von Thunen, J. 1826. *Der Isolierte Staat in Beziehung ant Landswirtschaft and Nationalekomie.* Hamburg.
Wachs, Martin. 1984. Autos, Transit, and the Sprawl of Los Angeles: The 1920s. *Journal of the American Planning Association 50,* 297–310.
Wachs, Martin. 1990. Regulating Traffic by Controlling Land Use: The Southern California Experience. *Transportation 16,* 241–256.
Wachs, Martin. 1993a. Learning from Los Angeles: Transport, Urban Form, and Air Quality. *Transportation 20,* 4: 329–354.
Wachs, Martin. 1993b. *The Role of Land Use Strategies for Improving Transportation and Air Quality.* Working Paper. Los Angeles: UCLA (October).
Wachs, Martin. 1994. Will Congestion Pricing Ever Be Adopted? *Access 4* (Spring): 15–19.
Wachs, Martin. 1995. The Political Context of Transportation Policy. Susan Hanson, ed., *The Geography of Travel,* 2nd ed. New York: Guilford Press, 269–286.
Wachs, Martin, and T. G. Kumagai. 1973. Physical Accessibility as a Social Indicator. *Socio-Economic Planning Science 7,* 437–456.
Wachs, Martin, Brian D. Taylor, Ned Levine, and Paul Ong. 1993. The Changing Commute: A Case-Study of the Jobs-Housing Relationship Over Time, *Urban Studies 30,* 10: 1711–1759.
Waldie, D. J. 1996. *Holy Land: A Suburban Memoir.* New York: Norton.
Walters, Alan. 1961. The Theory and Measurement of Private and Social Cost of Highway Congestion. *Econometrica 29* (October): 676–699.
Warren, Roxanne. 1998. *The Urban Oasis: Guideways and Greenways in the Human Environment.* New York: McGraw-Hill.
Wear, David (City Manager, La Mesa, Calif.). 1997. Personal interview. La Mesa, Calif. (April 21).
Webber, Melvin M. 1979. The BART Experience: What Have We Learned? Alan Altshuler, ed., *Current Issues in Transportation Policy.* Lexington, Mass.: Lexington Books, 195–134.
Weber, A. 1928. *Theory of the Location of Industries.* trans. C. J. Friedrich. Chicago: University of Chicago Press.
Weintraub, Daniel M. 1987. Assembly Vote Puts Brake on Orange County Toll Road Plan. *Los Angeles Times,* June 9: 3.
Weitzman, Martin L. 1974. Prices vs. Quantities, *Review of Economic Studies 41*: 477–491.
Wheaton, William L. C. 1959. Applications of Cost-Revenue Studies to Fringe Areas. *Journal of the American Institute of Planners 25,* 170–174.

White, Michelle J. 1988. Location Choice and Commuting Behavior in Cities with Decentralized Employment, *Journal of Urban Economics 24,* 129–152.

Willson, Richard W. 1992. Estimating the Travel and Parking Demand Effects of Employer-Paid Parking. *Regional Science and Urban Economics 22,* 133–145.

Willson, Richard W. 1995. Suburban Parking Requirements: A Tacit Policy for Automobile Use and Sprawl. *Journal of the American Planning Association 61,* 1: 29–42.

Witt, David (Director of Community Development, La Mesa, Cal.). 1995. Personal interview. La Mesa, Calif. (September 7).

Witt, David (Director of Community Development, La Mesa, Calif.). 1997. Personal interview. La Mesa, Calif. (April 21).

Wolf, Charles, Jr. 1993. *Markets or Governments: Choosing Between Imperfect Alternatives,* 2nd ed. Cambridge, Mass.: MIT Press.

Yen, S. T., and W. L. Adamowicz. 1994. Participation, Trip Frequency and Site Choice: A Multinomial-Poisson Hurdle Model of Recreation Demand. *Canadian Journal of Agricultural Economics 42,* 1: 65–76.

Zax, Jeffrey S. 1991. Compensation for Commutes in Labor and Housing Markets. *Journal of Urban Economics 30,* 2: 192–207.

Zax, Jeffrey S. 1994. When Is a Move a Migration? *Regional Science and Urban Economics 24,* 341–360.

Zax, Jeffrey S., and John F. Kain. 1991. Commutes, Quits, and Moves, *Journal of Urban Economics 29,* 153–165.

Index

access 17–18, 20, 63, 185n5, 190n1
 effect on mode choice 63–65
 pedestrian 49, 51–52, 64, 112–113, 189n6
 shopping 48
 transit 49
automobile emissions 17, 22–25
 air quality 22
 Clean Air Act (1970) 23–25
 components of 23
 in Los Angeles 24
 National Ambient Air Quality Standards23–24
 pollutants standards index 24
 Regulation XV in 25
 smog 22

Bay Area Rapid Transit (BART) 35, 120
behavioral framework 61
benefit-cost test 11, 173, 186n7
Bernick, Michael 18, 119–120, 162
Boarnet, Marlon G. 26, 28, 115, 144, 151, 162

Calthorpe, Peter 5, 8, 18, 39, 114, 150–151, 161, 185n4
Cervero, Robert 10, 18, 35, 48–49, 51–52, 119–120, 144, 162
choice framework 12, 58–59, 66, 75
Clean Air Act (1970) 21, 25
commute trips (*See* journey-to-work)
congestion
 high occupancy toll (HOT) lanes 29
 policy responses to 26–27
 pricing 21, 27–29, 177
 SR 91 28–29
 travel demand management in 27
Crane, Randall 34, 38, 115, 144, 151, 195n3

demand framework 54, 73, 172, 179
demand theory 57, 72
density of land use 29, 36, 38, 46, 48–50, 52–58, 68, 83
Downs, Anthony 8, 27, 196n13
Duany, Andres 5, 8–9, 18, 112

Environmental Protection Agency 23, 24
equity 17, 186n7, 195n4
Ewing, Reid 38, 50, 54–55
externalities 19–22, 26, 30, 175, 178

Fauth, Gary 55–56
Federal Transit Administration 29
Fulton, William 13, 112, 177

Giuliano, Genevieve 10, 33–35, 57, 144

Gómez-Ibáñez, Jose A. 20, 26
Gordon, Peter 26, 34, 56, 77, 89
grid street pattern 39–42, 45, 51, 64, 67–68, 70–71

Haagen-Smit, Arie J. 22
Hahn, Kenneth 22
Handy, Susan. 38–39, 48, 53–54, 64, 76, 114
high occupancy toll (HOT) lanes*See* congestion

Institute of Traffic Engineers (ITE) 9, 35, 112
Intermodal Surface Transportation Efficiency Act (ISTEA) 25

jobs-housing balance 10, 144, 190n8
journey-to-work 34, 36, 49, 55–57

Kain, John 17, 55–56
Kentlands, MD 8, 9
Kockelman, Kara 51, 57–58, 190n11

Laguna West, CA 8, 9
land development*See* zoning
land use
  mix(ing) 36, 40, 50, 52, 57, 74, 83–84, 87, 93, 98
  regulation(s) 13, 174, 195n7
Land Use Transportation Air Quality (LUTRAQ) 43–44
land use-transportation link 20, 25, 33, 36, 39, 57, 103, 110, 123
land use-transportation policies 18, 74
Levine, Jonathan 13, 107, 115, 119
livability 14, 29–30
  as transportation issue 18–19
  Federal Transit Administration "livable communities" initiative 29
Los Angeles, CA 22, 24
Los Angeles Metropolitan Transportation Authority (MTA) 125

market failure 111–112, 116, 195n4, 195n6
McNally, Michael 39, 49, 75
Metropolitan Transit Development Board 125, 147, 149, 155, 165
Mills, James R. 149

neighborhood types
  demand for 104
  regulation of 10, 112, 180
  supply of 79, 104, 110–111, 114, 116–117, 175
neotraditional plans/planning 6, 9, 45–46, 50, 63, 112, 171, 174
New Urbanism 5–10, 67, 114, 116, 171
  zoning and 112
non-work trips/travel 3, 9–10, 48, 57, 72–79, 84, 87, 93–94, 110–111, 172
North County Transit District (NCTD) 125–126, 197n7

Oakland, CA 46
Olmsted, Frederick Law 4, 35
Orange County Transportation Authority (OCTA) 125, 197n5
ozone 23–24

Pacific Electric 123
parking 21, 35, 43, 55, 151, 189n5
pedestrian pocket concept 5
planning directors, in San Diego 148–149
Plater-Zyberk, Elizabeth 5, 8, 18, 112
Portland, OR 9, 42–44, 51
price regulation (of externalities) 20–21, 178
price of travel/trips 22, 63, 69, 72–73, 77–80, 93, 103, 176, 178–179
Proposition 13 121, 137–138

quantity regulation (of externalities) 21

rail transit 56, 105, 120
  and Los Angeles 123

and Orange County 124
and right-of-way 132–133
and San Diego*See* San Diego Trolley
and southern California 125–127
station-area development 29–30, 114–116, 118–168
recreation demand literature 189n5
Regulation XV (South Coast Air Quality Management District) 25
residential location
market failures 111–112
preferences 109–111
regulation *See* zoning
travel behavior and 94–96, 102–104
Richardson, Harry 26, 34, 56, 89
Rosenbloom, Sandra 34, 91

San Diego
case study 147–168
TOD guidelines 151
Trolley 149–155
San Francisco, CA 3, 35, 45–46, 48, 51–52, 57, 120
Seaside, FL 5–6, 9, 70
Seattle, WA 46
self-selection bias 110
Shen, Qing 57
Small, Kenneth 20, 26, 34, 57
smog *See* auto emissions
social costs of travel 17, 19–20
regulatory solutions
price regulation 21
quantity regulation 21
transportation demand management 27
urban design and 18
social theory, in New Urbanism 8
South Coast Air Quality Management District 25
Southern California Regional Rail Authority (SCRRA) 125, 136
spatial mismatch 56
State Route 91 (SR91) 28–29
studies of urban form and travel
descriptive 44–47
multivariate 47–58
simulation 36–44

Taylor, Brian 27, 56
Texas Transportation Institute 25–26
traffic calming 68, 69, 71
traffic congestion 25–27
transit-oriented development 114–116, 119–124, 145–146
barriers to 144, 148, 162–166
Bay Area Rapid Transit (BART) 120
economics of 121–122
fiscal impacts of 165
history 122–124
in La Mesa, history of 148, 155–162
in San Diego 147–168
incentives toward 141–145
transit-oriented planning, models of 132–145
Transit Village Development Act (CA) 30
transportation demand modeling 43, 75–76
four-step method 4–5
integrated transportation and land use models 42
work trip mode choice models 55
Transportation Equity Act for the 21st Century (Tea21) 25
transportation planning, history of 17
transportation policy 13–14
in New Urbanism 8–10
urban design as 14
travel behavior
auto ownership 55
circulation patterns and 71
costs 62, 63
demand theory 57, 72
geographic scale and 76, 103–104
influences on
household income 50
personal attitudes 52
mode choice 63–65
nonwork travel 9–10, 75–76
pedestrian trips 53–54

travel behavior (*continued*)
preferences 62
regression models of 77–80
regulation of 21–22
residential density and 72–73
residential location 94–103
traffic calming 70–71
travel costs 73–74
urban design, effects of *See* urban design
travel diary data
Orange County/Los Angeles 80–84
San Diego 84–87
travel demand estimation
four-step method of 4
travel demand literature 55
travel demand management 27
trip generation*See* travel behavior

urban design
effects on nonwork travel 74–75
effects on travel behavior 74–80
travel costs 75
urban form
empirical analysis of travel 82–95
influence on travel cost 73–74
studies of 36–58
transportation influences on 33–35
urban location theory 33–34

vehicles mile traveled (VMT) 35, 38–40, 49–51, 68–72, 172

Wachs, Martin 5, 9, 71, 122
Walnut Creek, CA 45

zoning 112–116
benefits of 196n12
exclusionary 113, 195n8
fiscal 113, 121, 143, 195n8
measures of 126–132
public finance and 137–143
regression model of 132–137
versus actual land use 126–129, 192n9, 195n7